A Production of TimothyTuohy.com

Lean Hybrid Project Delivery

Copyright

Copyright 2021

© As "Lean Hybrid Project Delivery"

All rights reserved by Timothy Tuohy.

No part of this book may be reproduced or retransmitted in any form or by any means, graphic, electronic, or mechanical, including photocopies, scans, recording tape, or by any data storage and retrieval system, without the express written permission of Timothy Tuohy.

First Edition

For additional information:

Go to: http://www.timothytuohy.com

ISBN 978-1-365-67193-7

Dedication

This book is dedicated to my grandson. He was diagnosed with leukemia at the age of ten months, just as I began to write my Lean Integrated Project Delivery book. It was this seminal event that caused me dig more deeply into the science of project delivery.

The orange on the cover of the book is the result of my desire to call attention to Childhood Cancer and specifically to leukemia in children.

It is also dedicated to all those individuals who are out there managing projects. I hope this method will help you.

Lean Hybrid Project Delivery

Contents

Preface 3
1 – Introduction 7
2 – Basic Keys of Success 14
3 – Conditions of Satisfaction 27
4 - Charette 40
5 – Flesh-out 50
6 - Sequencing 70
7 - Commitments 83
8 – Monitor and Control 107
9 – Completing the Steps 126
10 - Closing the Project 136
11 – Visually Speaking 138
Appendices 145
    The Five Lean Principles 145
    Manifesto for Agile Software Development 146
    The XP Values 147
    Meeting Tips 148
References 149

Lean Hybrid Project Delivery

Lean Hybrid Project Delivery

Preface

Welcome to this adaptation and application of Lean Hybrid Project Delivery – I promise this will be a turning point in how project delivery is viewed. The methods herein are all proven effective processes. Before you continue answer begin thinking about these four cornerstones of your success:
1. Commit to constant improvement.
2. Clearly define your desired outcome.
3. This book provides the best tools, study them and learn them.
4. This book provides the best methods and processes to follow through, study them and learn them.

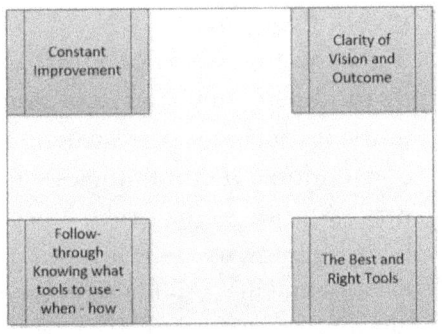

Having just completed a two-billion-dollar construction program that built the most advanced rehabilitation center in the Southeastern United States, a new one hundred bed hospital, and a three-story vertical expansion that provided fifty-four Intensive Care Unit rooms, I decided to share what I have learned. This book will share those lessons with you.

It is not because I think myself a Project Management genius that I write this book. I find myself pursuing constant improvement. I have found tools that work and want to share this with everyone who needs tools that work. As I studied the topic, and produced papers on the subject matter, but these have changed as I have studied, learned more, and applied the knowledge to the methodology. Maybe what I am learning, and have learned, can

Lean Hybrid Project Delivery

be of use to you.

If it is, maybe you will share this with your peers. As a team, maybe we can perfect this method of Lean Hybrid Project Delivery and all of us can benefit.

> Do something great ... you get used to being famous ... you never get used to being ignored.
>
> – Patrick Winston, MIT

I work diligently in this book to exclude people from the term "resources". This methodology though, is not specific to Information Technology, nor is it specific to Construction. Planning projects in the conventional methodologies where one attempts to address the process prior to the process beginning is inherently flawed because humans are notoriously incapable of seeing the future clearly. Further, Agile methods don't apply themselves well because usually we are not producing original coding and have no control over the segments or sprints. This then, is a *hybrid* solution.

I have also been integrating and adapting Lean, Kanban and Agile methodologies into the Lean Project Integration methods described herein. We too are learning this and adapting as we apply it.

I wish to express my gratitude to William R. (Bill) Seed for introducing me to Lean Construction Institute's Last Planner System. The book "Transforming Design and Construction – A Framework for Change" for which he served as Executive Editor, has been the major inspiration.

Lean Hybrid Project Delivery is not a complete treatise on the

Lean Hybrid Project Delivery

subject matter and is not intended to be. There is a list of books and authors that have formed the reference material for this content in the References at the end of the book.

For the last six years I have been working together with a cross functional team of project managers, designers, architects, and construction engineers. This book follows and earlier book that started with a Power Point training. I will use the APA model for reference to the books, articles, web sites, and journals referenced herein. I recommend you follow-through by "going deeper" and expanding your knowledge as you need it and as we develop our skills. This book will give you the basics so you can begin on your own course of action.

The vision for this book is to provide the information and methodology to help you succeed in implementing this system, so we can stop managing strategic projects and start delivering them successfully in a Hybrid, Lean, Agile, Kanban and Predictive integrated methodologies.

Throughout this book there will be comment boxes relating to the most important yet least studied topic necessary for success, emotional intelligence. This was important enough that Cornell University placed it as the first of the classes in the Project Leadership course I took at the end of 2020 during the global COVID-19 pandemic. The pandemic raised the awareness of mental health in the professional workplace. The understanding of emotional intelligence as a study unto itself has been

---
THINK

If you want to be emotionally fit …
get out of yourself.
-- Tony Robbins

Lean Hybrid Project Delivery

something largely overlooked in the profession of project management. Indeed, it is not part of any of the project management courses I have taken through the years. It is this addition to our studies, more than any other topic, that moves this book into the field of strategic project leadership.

This book will purvey a singular focus relating to the projects we take on in the future, that to succeed, we must invest of ourselves in the people on the teams.

Tim

Lean Hybrid Project Delivery

1 – Introduction

This book makes two assumptions. First that you are already managing projects and know what you are doing, or second that you are a customer of Project Managers, and are affected by Project Management Technology, and (in both cases) are dissatisfied with the approaches to Project Management you have been using. This is the culmination of five years or more of deep immersion into strategic projects. While I considered myself a good project manager for many years before the opportunity to participate in these strategic projects presented itself, it wasn't until I began this adventure that I really understood. It was with these grand strategic projects that I began to learn the subtle nuances and advances in project delivery that could be best translated as "hybrid". (Flávio Copola Azenha, 2021)

If failure is not an option ... if your project is strategic or disruptive then this methodology is likely exactly what you have been searching for.

This book includes a methodology I was introduced to through construction, but it also includes adaptations of Lean, Agile, Scrum, and Extreme Programming as well as (where appropriate) predictive or waterfall methods. I have studied, applied in action, and adapted these methods. Project delivery has evolved to a new level of Lean Hybrid Project Delivery.

It would be remiss of me to not give credit where credit is due. When first I began doing this massive project, I would have done it in a standard predictive (waterfall) methodology and probably would have been unsuccessful. At least, not as successful as I have been.

Having been a student of W. Edwards Deming's methods since the late 1990's, I consider myself a Lean practitioner. Lean having

Lean Hybrid Project Delivery

started with Henry Ford in 1913. Deming is credited, for example, with helping to steer Ford Motor Co. toward building cars with fewer factory defects beginning in 1981. (Walton, The Deming Management Method, 1986) The term "Lean" was coined by a graduate student at MIT in the 1970's and 'stuck'. (Krafcik, 1988) We will spend some time with Deming's Cycle in this book as it is one of our foundational elements.

Also, I give substantial credit to the modification of my thought process to William (Bill) Seed. It was Bill Seed who introduced me to the Last Planner System which opened a

> A leader is a coach, not a judge.
> W. Edwards Deming

path to lean agile project delivery from the construction profession. I don't think he referred to it or understood it to be anything near what agile is or even cared what agile was as he was a construction man and never in Information Technology. Still, I owe him a deep debt of gratitude for being the one to expose me to that process. It was a result of that exposure that I wrote my 2017 book on Integrated Project Delivery. (Tuohy, 2017)

The strategic project I will be using as a reference in this book was a group of five projects but consolidated as strategic projects for Jackson Health System in Miami FL; these constituted roughly $2 billion worth of construction in the South Florida specifically in the Miami area. This constituted the replacement of a Rehab Hospital that was very dated with a new nine story state of the art Rehabilitation Center. It included a newly constructed full service 100 bed hospital. It included a three-story vertical expansion.

There's quite a story behind the three-story vertical expansion. First, there was to be a six-story building that that would house a massive expansion of the Emergency Department, observation rooms, two floors of Intensive Care rooms, and three floors of

Lean Hybrid Project Delivery

specialty in patient rooms for the Transplant Institute. However, this building was struck because of financial constraints and was substituted for a two-story Super Walmart sized Emergency Department without the bed tower. That too was struck for financial constraints. The three-story vertical expansion was proposed to be built on the existing Ira Clark Diagnostic Treatment Center. This was the final iteration. The three-story vertical expansion was finished and opened. It included fifty-four state of the art Intensive Care Unit rooms. In the appendix you will find a letter from the Senior Director of Clinical Systems in IT that we were able to open that on March 2nd of 2021 with zero issues. I'm calling that to the reader's attention because it demonstrates the depth of the capability of this hybrid project management methodology that I am talking about in this book.

These projects also included renovations of all floors in the six hospitals that form the core of Jackson Health System. Additionally, Jackson Health System opened Urgent Care Centers around the County.

In this book I want to focus on the strategic or capital projects. For this book Strategic Project is defined as those projects that happen in a business or organization that add or change substantially how that business or organization might be operated. In this case adding two hospitals and adding the vertical expansion substantially changed how that health system functioned as a matter of their daily business. Therefore, I would refer to these as strategic projects that substantially changed how the business or organization does their business on a day-to-day basis operationally.

A strategic project could be a moon shot for the purposes of establishing a colony ... that would substantially change the way a government did business. A strategic project could be a project to build a new video game, if you are a video game business and you have one primary massive multifunction Internet based software

Lean Hybrid Project Delivery

game the people who are playing are the core of your business. If you are adding another game this is a strategic project because you're making a strategic change in your business and there are substantial risks. If you are an energy company who has defined a way to manage and supply energy without polluting the air or the ground or anything else and thereby are changing the way humanity is able to live ... that's a major strategic project. Alternatively, (and more practically) a building project can be a strategic project. Most capital, mega-projects can be considered strategic projects.

This book will open the doors to individuals who want to be able to manage such projects. It is important to understand that this book is not an exhaustive treatise on the matter of Strategic Project Management, or Lean Hybrid Project Delivery, and it's not intended to be. The format of this book will be such that the processes, procedures, and activities that I developed during the projects of the last six years will be interlaced between them on every chapter. During the process of this six-year adventure, I have been taking courses at Cornell University and MIT for project management and artificial intelligence. I will be weaving some of those lessons into this book as well. (Goleman, 1995) Throughout each chapter I will be putting in different sections that break the processes and the procedures down and gives the underlying leadership functions that must also be employed.

The reason for this is, in many cases, I have found that the project management books tend to separate out the processes, procedures, activities and leadership functions into separate sections in separate chapters where they are very interlaced in the day-to-day workflow. This book is designed to be more workflow oriented than any of the books that I have been familiar with so far. We will endeavor, as you the reader and I, the writer, go through this book, to talk about the emotional intelligence that is required for leadership. Emotional intelligence is something that I have had no education on until I

Lean Hybrid Project Delivery

started at Cornell. At that time, I began to understand how important it is for the strategic project manager to have deeply understood and deeply grasped emotional intelligence principles of leadership. Projects managed in the conventional manner of working to produce a plan for something to occur in the future through the four steps; Planning, Build-up, Implementation, and Close Out (Harvard Business Review, 2012) simply don't work. We apply these methods because we need tools and there are management demands for status and progress reports, but the project manager is relegated to being a 'box checker' at best and a task master, similar to what I imagine were employed in the construction of the pyramids, at worst.

Projects managed in the agile methodology generally are not applicable. Even though the rapid, iterative application of installing and integrating components and modules of the Electronic Medical Record or other major software system may make one 'feel' like an agile methodology might work. The Agile Manifesto (Agile Alliance, 2017) and Twelve Principles are organized specifically for developing and delivering software. Most Enterprise Information Technology organizations are not developing software but are implementing someone else's software and either customizing it to their organization or forcing their organization to comply with the original structure. In the case of the Health System for which I work that Electronic Medical Record and the associated systems is provided by Cerner. (Cerner, 2017) Because of this, agile methodology is not applicable in a meaningful way in the Healthcare Information Technology environment, at least, not in the one I was associated with.

Projects managed in construction apparently experienced similar problems. Like Information Technology, the construction industry was seeing projects exceed the planned time and cost, as well as creeping scopes and customer dissatisfaction. Their solution was to look to the methodologies that are defined by the Lean Principles

Lean Hybrid Project Delivery

that have grown up through the years from 1913 when Henry Ford implemented the first Lean model to his entire production line. No one called it Lean back then but that is what it was. More recently Kiichiro Toyoda and Taiichi Ohno shifted the focus at Toyota after World War II to the model widely known today as the Toyota Production System (Liker, 2012). Lean techniques were already in practice by Eiji Toyoda and Taiichi Ohno when the end of World War II thrust upon Toyota postwar Japan's fertile environment for new thinking and W. Edwards Deming's quality enhancing ideas were adopted. (Womack, 2007)

Projects managed in the Lean Construction Institute's version of Integrated Project Delivery introduced the construction industry to a new version of project management. Through the years from 1996 until 2007 there has been a series of progressive iterations and steps that have led up to the codification of Target Value Design (Ballard I. D., 2016) by the Lean Construction Institute.

All of these impacted my need to learn and develop a more perfect methodology of project delivery. I was not alone. Since writing my 2017 book and studying at Cornell, I have learned there is a general movement toward a Hybrid methodology. This book encapsulates that methodology into Lean Hybrid Project Delivery.

This book is designed to walk you through the steps quickly and easily; a sort of "**Vade Mecum**". Vade mecum is Latin for "go with me." In English it has been used since the 1600's as a term for manuals or guidebooks sufficiently compact to be carried in a deep pocket. But from the beginning, it has also been used for such constant companions as gold, medications, and memorized gems of wisdom. (Webster Dictionary, 2017) "Go with me" is very much a methodology of the Toyota Way and therefore fits the spirit of this methodology, it is as valuable as gold, and as important as your medications.

Lean Hybrid Project Delivery

It is not too early to start learning the system. Don't worry about doing it correctly right from the start. Lean Learning is about continuous improvement!

> The genesis of Lean methodology stems from Toyota Production which was inspired by W. Edwards Deming and others.
>
> Bill Seed

In writing this book, I wanted to help you dive into the processes and be successful. I also wanted to provide enough references that if you really wanted to dig in, you could.

Finally, and in closing, I encourage you to read the reference material that I will refer to in this APA formatted book. If you read something herein, please do not take my word for it. Please do go and read these documents, books, reports and Journal articles so that you too can gather a depth of understanding of how these lessons apply to what we do when we are working to make strategic changes and execute strategic projects in our organizations.

Ready? Let's begin.

Lean Hybrid Project Delivery

2 – Basic Keys of Success

> **It's a mindset**

Basic Keys of Success
While there are methods to Lean, Lean is a mindset, a set of principles ... not a set of rules, not a methodology. Agile development is a process for rapid software delivery that is connected to many Lean principles. While many capital projects include Agile elements, Agile is not suitable for the entirety of the project. Here we adapt Lean, Agile, Scrum, XP, and Kanban to any project, even if the upper level of project is predictive or waterfall.

Executive Ownership and Sponsorship
The largest and most complex projects (Strategic Projects) command budgets exceeding $1 billion and require three to five years (or more) for design, planning, and construction. The magnitude of such projects creates distinctive complexities - multiple interfaces with stakeholders such as local communities and government bodies, new regulatory and environmental requirements, and often unique technological challenges. In this book I define projects of this scale as strategic. A strategic project by its very definition is an executive level project therefore it is assumed that there will either be executive buy in because this project was proposed and adopted by the executives or is an executive originated project. The executives take full ownership of the outcomes. It is the responsibility of the Lean Hybrid Project Delivery Professional to make sure the executives are well informed so they can step in and make the tough decisions quickly when needed. (Sergey Asvadurov, 2017) In the case of construction of a new hospital or a new warehouse or a new plant or a new Starship ... the orders to do this, come from the highest levels. The

Lean Hybrid Project Delivery

executive branch (that is the CEO) will often be reporting to a board of directors. The CEO may the Chairman of the Board, but there will be executive buy-in for the capital project. The funding will have come or been defined prior to the strategic Lean Integrate Project Delivery Professional's direct involvement. It would be best if the strategic Lean Hybrid Project Delivery Professional is involved at the beginning of discussions for the project when the board is determining whether or how much the funding for that project should be. Thereby, the strategic Lean Hybrid Project Delivery Professional has the ability to understand what the intent was for that project from the beginning.

We will get more in depth on this in the next chapter as we determine how we are going to define what is going to be done in the project and the Conditions of Satisfaction Meetings. There are series of meetings that will occur during the process of the strategic project, these include the planning, execution, Delivering, and Closing of the project. While these are somewhat different than the PMI terms, I will tie them together in this book.

> "If you can't explain it simply, you don't understand it well enough."
> Albert Einstein

Deep Collaboration

Omnino Collaborare - labor together. Latin from *com* + *laborare*.

Collaborate – REALLY COLLABORATE with the objective of truly, constantly, and incrementally improving the process and delivery of projects. Most Lean process promoting writers tend toward the Japanese words because so much is based on the work done by Eiji Toyoda through the years. An American self-help coach, Anthony Robbins, coined a word "CANI!" – really an acronym for "Constant And Never-Ending Improvement!" so he wouldn't have to use the Japanese words "Kai Zen" which mean change and good. Together

Lean Hybrid Project Delivery

"kaizen" is the term used frequently to refer to the process of finding and making small incremental improvements in pursuit of perfection.

I like to use Latin because it so echoes back to all those great military statements of courage, valor, and commitment, in the case of continuous improvement, Latin tends toward "semper melius." I like it better, but most Lean thinkers prefer to honor the Japanese words.

> Semper Melius or Kaizen ... The Japanese word is widely used and may be easier to remember – but I like the Latin.

What is it about "semper melius" that applies itself to this topic? Collaboration, should include all the stakeholders of the project, is a systematic study in pursuit of the perfect delivery of the project's product or outcome.

What does this mean? Collaborate – really collaborate. This means as Project Managers we will need to broaden our horizons. We need to include not only the Information Technology Department, but as in the case of the Healthcare profession, nurses, physicians, programmers, technical support, IT infrastructure, marketing, and staff but also the end users. Our goal is to bring all the parties together in collaboration from the beginning to establish true working understanding of the needs of the Health System or business that we as Project Managers are trying to fulfill.

Why Collaborate? Without deep collaboration with all the associated people related to the project, there are hurricane category winds that will blow us off course seemingly on a whim. By engaging deeply with all those associated with the delivery of the project, and allowing them to own it, we as the Lean Hybrid Project Delivery Professionals are no longer the 'task master'. We have transitioned to Project Leader. It

Lean Hybrid Project Delivery

is a very agile concept for a scrum master to be the one who knocks down obstacles for those people who are occupied producing working software; in Lean Hybrid Project Delivery methodology a similar transition occurs. It is at once liberating and fulfilling for the project team.

Philosophy and Best Practice

Managing Projects in Healthcare (for example) has not been considered as a process that can add value to the total beyond the thought "the sooner the software or product is in service the better we can perform patient care." This is because the sooner we can get the electronic medical record to be the single source of truth for the patient the more likely we are to provide better informed care.

There are important financial and legal reason why project managers need to find a more effective, efficient, and expedient way to successfully deliver projects. The success of the entire business depends on it. The primary purpose for collaboration across all the parties both providing and affected by the project, is to establish the definition of "true north" as it applies to the project being delivered. (Lean Enterprise Institute, 2017) Having a "true north" allows us all – as a team – to come back to the right direction to accomplish our goal. If the path we are taking is not true north, then we can quickly and easily adjust to the course that will take us in the direction of true north.

> The largest and most complex projects (Strategic Projects) command budgets exceeding $1 billion and require three to five years (or more) for design, planning, and construction.
> McKinsey and Company

There is no perfect project management style. There is no perfect project management method. What we can offer in this book is

Lean Hybrid Project Delivery

proven methods. We may not be perfect and will require constant adjustments, but we provide proven methods that deliver projects successfully.

Keep trying

Applying Lean Principles of constant improvement also means always looking for a Plan B. Have multiple plan B's. Always be looking for an alternative to reaching your goal. I once had a friend who was a pilot of private airplanes, he told me, "I am always looking for a place to crash and still be able walk away." This principle completely applies to this item of our list. Thomas Edison said, never get discouraged if you fail learn from it and keep trying.

> Never get discouraged if you fail. Learn from it. Keep trying.
> Thomas Edison

Treat Failures and Delays as Learning Experiences

Treating failures and delays and missteps as learning experiences does not imply accepting running delays as commonplace. This is very important, if an individual makes a mistake there is a need for them to feel confident in owning it immediately and allowing the team to compensate and improve. Everyone who's showing up at work (with very few exceptions) is doing so with the hope that what they are doing will have value, that they will add value, that they will come away at the end of the day feeling like they have done good. No one wants to come home at the end of the day and see their family and feel as if their life is just a waste. We need to (as project leaders) assume that everyone who's showing up for work is doing so because their work adds value to their life. Their work adds meaning to their life, and they want to do the best job that they possibly can. For those of us who are constantly studying, learning, and going to school (and spending money on our education) it is sometimes difficult to assimilate that people who are tradesmen, or are labor feel like it's

Lean Hybrid Project Delivery

important for them to do the best job they can for their own reasons. We as a team must always understand that every team member is doing their best with what they've been given, and they want to succeed as part of the team. Therefore, if the individuals are not performing as well as they should or if things have been missed or are out of their control, they need to feel free to be able to learn from this and we, as a team, need to improve by it. We'll spend a lot more time on this throughout this book.

We value individuals and interactions, over processes and tools. (Agile Alliance, 2017)

Accent Small (and Large Successes)
In this book we will be accenting the need to celebrate small and large successes. Ellon Musk said in an interview, "Life cannot just be about solving one miserable problem after another ... that can't be the only thing. There needs to be things that inspire you ... that make you glad to wake up in the morning and be a part of humanity." Similarly, everyone who's working in the project needs to feel recognized and we will apply Maslow's Hierarchy of Needs (Maslow, 1943) as we talk about the internal need of every human being to feel like what they're doing is important, that their life has meaning, that their job has value, and that they are participating in something that's important to people. In a strategic project this is so very important. If it's not an important project, it's not strategic. Everyone who is participating in a strategic project needs to feel some ownership in the delivery of that project, otherwise that project will not be delivered successfully. It's your job to make your team members famous.

There is a Method for Success
There is a process for reaching the successful delivery of the project. The follow through discussed as the fourth cornerstone of success is mapped out and there are associated meetings. Every meeting must by its own nature be a "safe zone". For this methodology, to be

valid, functional, and successful, every participant in the deep collaborative methodology (or working group) must feel safe to express their goals, ambitions, restrictions, inhibitions, and fears. If any member of the team feels insecure as a member of the team, they are likely to keep their mouth shut withhold information to protect themselves. This is detrimental to the success of the team. It is very important that all team members have a feeling of security as they move forward through the process of becoming team members. However, this comes at the cost of trust. Team members earn that trust by commitments made and kept to the team.

This also means that as the project leader, one should maintain a status of protecting all the members. If one feels that an individual member is not participating at the level that is needed for success, the conversation is held with that individual as soon as possible to assure that that individual is not withdrawing for an unreasonable excuse.

Use a Big Room Approach
For this chapter, we want to focus on some basics to successfully building collaboration teams. We have been working with this methodology for several years, and though I am still a student, this works. You might have picked up on that by the construction of the 'chapters' of this book. You'll note that each chapter is written much like a school paper, broken down into sections and providing references to the material. From this you can derive that I am studying and practicing it in the lab of real-life applications.

Concentrate on Flow
Lean is not just the elimination of waste; it is also focused on flow. This methodology focuses on the flow of work through the project, through the teams, through the phases, to the successful delivery. It is important to look at how the various parts of the project link together, where the hand-offs are, and how these touch points work with each other.

Lean Hybrid Project Delivery

Follow the Process

There are a series of steps listed below. These steps are required, they are the "Science", the mechanics of success. Success keys one through eight address the "ART" of Lean Hybrid Project Delivery, the following Success Keys in their various parts are the "Applied Science" of this methodology.

The Science Keys are as follows:

1. **The Conditions of Satisfaction.** This is a meeting, or group of meetings, depending on the size of the project. In this meeting we will be talking about how to build a list of all the items that the CEO and other chief executives are looking for at the completion of the project. This is a crucial meeting whose outcome defines "<u>done</u>". The "Definition of Done" (Alliance, Definition of Done, 2021) is the most important and final milestone of the project. Without this at the beginning of the project you do not know what is going to happen, what you need to do to get the specific measurable end result that is acceptable and satisfies the requirements. This is a complete analysis of the finished business environment for the project. This provides the Clarity of Vision or True North.

2. **The Charette.** A Charette is an intense planning session. This is a nod to the architectural profession as Charette was a two wheeled cart that was used pre 1900 in Ecole des Beaux Arts in France. It is said that architectural students were given a design problem to solve within an allotted time. When that time was up, the students would rush their drawings to the studio in Ecole in a cart called a Charette. Students often jumped into the cart to finish the drawings on the way. The term evolved to refer to the intense design exercise itself. Today, it refers to a creative process akin to the visual brainstorming that is used by design professionals to develop solutions to design problems within the

Lean Hybrid Project Delivery

limited time frame. (Gail Lindsey, 2009) I like this term so I'm going to use it.

I'm going to use it to define the brainstorming (Osborn, 1953) session that happens next in the process of planning. It is best when the chief executives are in participation in the Charette. (It is very helpful.) However, this can be a meeting for the next tier of management, executive vice presidents, senior vice presidents, vice presidents, and senior directors ... that level of individuals who are charged with the execution of the project and the assurance that the project stays within the budget. As everyone in the corporation reports into their structure at some level, the importance of this meeting cannot be overstated. This is where we define the milestones. This is where we define the high-level deliverables necessary to meet the Conditions of Satisfaction.

3. **The Flesh Out.** This is where we take those high-level activities and milestones that were defined in the *Charette* by the executive officers, vice presidents, and senior managers and start bringing that down into what needs to be done to finish those processes. This meeting is attended by the directors and the line managers of the organization. Here the team is building the activities that need to be done to finish the project and deliver it on time. There are several processes that are involved in this, and they are outlined in the chapter that is referred to as Flesh Out (Merriam-Webster, 2021). This is an extensive process there are several different planning methodologies that I will put in the chapter to explain how to do this. This is all iterative and extremely collaborative.

4. **Sequencing.** After this, we have a series of sequencing meetings. Mind you all of these meetings; the Conditions of Satisfaction, the Charette, the Flesh Out and the Sequencing meeting are all done at a Kanban Board (Alliance, agilealliance.org/glossary/kanban, 2021) so it's very important

Lean Hybrid Project Delivery

that one has a big room in order to put the Kanban out in front of everybody. The cards need to be where they can be seen. Kanban means – a card you can see. It's important the team can work with the cards, there are several methodologies to do this. We have learned in 2020 and 2021 how to do this in a pandemic restricted environment. However, it is much easier to do this person to person, face to face, so that as we are talking about the various ways of doing things, we can begin to move the cards around into the correct sequence and make sure that we have identified the constraints and the prerequisites, in the correct order. We'll talk a more about that in the Sequencing chapter. This is all iterative and collaborative.

5. **Commitments.** Now we start talking about commitments. What we've been doing through the last three segments of planning is building a "network of commitments", each of these cards that has an activity or action on it have been reduced to the smallest possible time frame (called chunking down) that could be associated with this individual work item or activity or task that can be assigned to an individual or a small group. Now we are moving those cards (that you can see) from the Kanban board into a spreadsheet where we can begin tracking them in date order of when they should be getting done and how we can do it.

Tracking commitments and working with the team members has two favorable outcomes. First, it helps reduce stress by making the individual work items more manageable and second, it helps build trust. (Robert Conti, 2006) This is all iterative, occurring weekly or biweekly and extremely collaborative.

6. **Monitoring and Controlling the Project.** There will be a chapter on monitoring and controlling but it will be focusing on reporting and the process of the status meetings and how to do the things that need to be done. It would be inaccurate to think of this chapter as being a standalone item. Monitor and control is

Lean Hybrid Project Delivery

matrix throughout the entire project. You will see in this book's chapters "text boxes" that give the monitoring and controlling processes that apply during that chapter's subject matter. Planning and executing this project will include references to monitoring and controlling methods interlaced into the various functions that are necessary.

This method focuses on the final Business Environment from the beginning of the project.

7. **Completing the Steps.** Then we'll talk about completing the project. There are several processes that we will be performing throughout the life of the entire project to assure that things are efficiently and effectively completing and that they are being completed to the quality needed. The effectiveness of things being completed will be validated by a series of processes and tools, among these things is a punch list. One of the things that I've learned over the last six years delivering $2 billion strategic projects is, checklists and punch lists are needed as you near the end of the project. It is the same whether you're dealing with IT or with construction and the trades. When you near the end of the project, before you start buttoning everything up, you need to be passing through and doing punch lists to assure that things are happening where they're supposed to be happening, when they're supposed to be happening, and that they look like what they're supposed to look like. (Anita Tucker, 2013) This is called Management by Walking Around, it is a criterion for success, and we'll talk much more about that in this chapter.

Lean Hybrid Project Delivery

8. **Delivering and Closing the Project.** Finally, we'll talk about delivering the project. There are so many things that are involved in delivering the project that people just overlook. One of the things that makes projects successful is (as you near the end) you know what it's supposed to look like ... but does the person who is going to be using it know what it's going to look like? Do they know how they're going to move into it? Do they know how they're going to process it? Do they know how things are going to happen once it goes into operations? Part of the process of delivering the project is helping the people who will be using the project's product when it's done to understand and get to know this process. I call it a process of "dry runs", this includes rehearsals to ensure deployment success and also includes planning and rehearsal of rollback scenarios. In the case of a new facility, particularly where you're hiring new employees to do the job or jobs that are required to make that facility work, not only do you need to understand what those jobs are, you also need the employees and managers to know where those jobs are, what those jobs are, and how to do the jobs. It's important that if you're going to open a new warehouse (I just walked past an Amazon box so I'm going to use an automated warehouse as an example) if you're going to open a new highly automated warehouse that will enable you to serve an entirely new market, you have to know how it is all going to work,

> The planning fallacy refers to a prediction phenomenon, all too familiar to many, wherein people underestimate the time it will take to complete a future task, despite knowledge that previous tasks have generally taken longer than planned.
>
> Kahneman and Tversky

Lean Hybrid Project Delivery

how does it all work together, how do these boxes come through, who's touching them, do they get touched, where do they go, how does it get to the truck, and all those new employees need to be trained on that method that's part of delivering a project.

The final business environment is a conscious focus of the project from the beginning. It is part of the Conditions of Satisfaction. Closing the project means successfully providing the necessary business environment and turning the project's product over to the owners for their operation.

Finally, there is all that part of the project that closes all the legal and financial elements.

What does this do for us?
1. It provides a concise path to leaner, less wasteful process decisions.
2. It supports project process and methodology tailoring.
3. It clarifies process and methodology for easier identification of appropriate strategies.
4. It takes the guesswork out of performing the project work for successful delivery.
5. It helps clarify risks and therefore enables the likelihood of success.

These are just some of the fundamental advantages of employing this methodology of goal-driven approaches. (Mark Lines, 2020)

Lean Hybrid Project Delivery

3 – Conditions of Satisfaction

> **It's your vision**

Conditions of Satisfaction
It's the vision of the leadership that commissioned the project. This process identifies the specific, measurable purpose of the project, based on that vision, adding clarity, definition, and purpose to the outcome.

This definition includes how it is attained, assures it is reachable and sets the timetable for its attainment.

Lean Hybrid Project Delivery is a transformational change.

The "Conditions of Satisfaction" is the single most important part of the project. This is the meeting with the executives that clearly establishes and defines the final milestone of the project. You may have heard this referred to as the "Definition of Done" (Alliance, Definition of Done, 2021). Up until now, the outcome from this type of a meeting has been a Project Charter (Hayes, 2000). A Project Charter may still be an outcome of this meeting if it is needed. However, the conventional Project Charter as well as the conventional Vision Statement may soon have a different comprehensive definition as we establish the Conditions of Satisfaction. This is not to say that the Vision Statement is obsolete, it is not. However, by defining the Conditions of Satisfaction we expand on and clarify the vision for the project. The Vision Statement then, can be thought of as an outcome from this meeting, a summary statement of the Conditions of Satisfaction. These are the complete definition of the final milestone.

Lean Hybrid Project Delivery

During the process of the Kanban meetings that final milestone is just a diamond on the wall that says "in operation" for instance. I worked at a healthcare system whose motto was "Miracles Made Daily", on the final milestone the words "Make Miracles" summed up the Conditions of Satisfaction.

While the Conditions of Satisfaction is greater than the outcome of the meeting's list of criteria, it is developed as a bulleted list of the specific measurable results that will satisfy the executive staff that their vision of the project has been reached.

As a team we are looking to completely understand and design the finished and operational business environment. From this point in the project to the end, every step of this methodology is focused on delivering the finished business environment.

To begin the bulleted list, I will ask questions like, "what is the specific measurable outcome you wish to have as a result of this project?" I might also ask, "why are we doing this project?" We want to identify the specific purpose that is driving this project to its conclusion. These questions can be answered with things like, "market research shows that our customers want this". It may be a question of patient satisfaction in

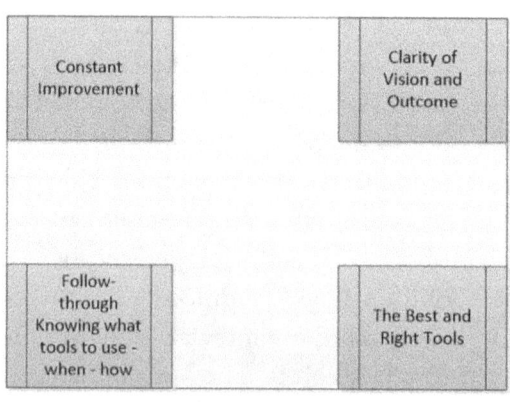

the healthcare environment, or it may be a new building or new disruptive technology that your company may want to purvey. Next to the cornerstone of "constant improvement" this cornerstone, "clarity of vision and outcome" i.e., knowing the

Lean Hybrid Project Delivery

business environment outcome, is the most important cornerstone in this project methodology. It is responsible for keeping everyone on course. It points the way to true north. It is the target for which everyone is working.

The importance of this cannot be overstated. Let's use the example of building a new facility to house intensive care unit rooms. Among the things that we want to address in the Conditions of Satisfaction is what to leave in and what to leave out. We will ask ourselves what we are happy with in the current state environment and what are we unhappy with. It's critically important as we move forward to completely understand the business environment that will be exercised in this new project. I'm going to be referring a lot to new buildings, but this applies to major rollouts of software, whole new business processes, and any number of other strategic projects that we may be applying this methodology to. It's as important, to the successful outcome of the project, that we clearly understand what we don't want as much as what we do want. It's critical to understand what we're excluding from the project as much as what we are including in the project. Defining this cornerstone and the "constant improvement" cornerstone are focused on what is termed "soft organizational and leadership elements" of project delivery. This is the "art of project leadership". (Sergey Asvadurov, 2017)

The other two cornerstones of success in this methodology are "the best and right tools" and "follow through – knowing what tools to use, when to use them and how to use them". These two employ methodologies from the multiple project management bodies of knowledge, they are the "science of this project delivery" in this methodology. There is an unstated mortar between these corner stones, it is communication.

Lean Hybrid Project Delivery

In this part of the hybrid methodology, the Conditions of Satisfaction encompasses the primary methods of communication, particularly if we are working in a distributed remote setting. It should be clarified which protocols of communication will be used and how they will be used. Is there an in-house standards or software? Perhaps we will use Sharepoint or some other shared platform that employs a "check-out / check-in" technology to assure documents are updated serially. Will there be a shared drive set of folders that will be used for the project? What will the naming convention of these folders and the documents stored therein be? Who has access to these folders and what level of rights do they have to read, modify, edit, etc.? What will the communications standard be? Will we use text messaging? If we are using email what will the standard be for the subject line? How will status be reported is there a standard, or will we use the Lean A3 methodology? Who is included in communications is defined by layers? An executive

> Keep in mind communication in a virtual world can be very challenging. Many team members are feeling stress from multiple sources. Surfeit and stochastic are two words that it seemed to purvey the new standard operating procedure for integrated project delivery in a virtual world with everyone trying to keep things from career to homelife from collapsing.

Lean Hybrid Project Delivery

does not need to be deluged by daily emails between team members relating to coding issues or other daily activity discussions. Communications are crucial to project delivery success. This is outlined and agreed to in the Conditions of Success at a high level primarily to gather what communications the executives and primary stakeholders are expecting and documenting that to their satisfaction.

There is another important aspect of communication. Communication is the foundation of building teams. Having the basis of communication agreed to at the leadership level sets the expectation for the entire project and its teams. The leadership here must also embrace the mindset of Agile (and Lean) that empowers the team members to accomplish the delivery of the project in a self-organizing method. We will discuss and agree to the methods for:

> Well, those drifter's days are past me now - I've got so much more to think about - Deadlines and commitments - What to leave in, what to leave out
>
> Bob Segar – Against the Wind

- Maintaining focus on a shared goal
- Delegating authority appropriately
- Watching for ways to leverage team members' strengths
- Watching for people's weaknesses and making adjustments for them
- Giving each person encouragement, empowerment, or direction at the right time

As we are constructing the Conditions of Satisfaction, we'll want to look at high level gap analysis as well. Unless this is a brand-new product, service, or facility, this project is likely to be

Lean Hybrid Project Delivery

changing an existing product, service, or facility. It may be changing all three, as in a facility renovation. The Conditions of Satisfaction should also include what needs to be eliminated or changed to satisfy the executives and primary stakeholders.

This brings to mind a Lean analogy. The story of Grandma's ham. The story goes like this:

Once there was a man who married his high school sweetheart. They married in the spring and when Thanksgiving came and, among the other items purchased for the dinner, was a fine large ham. Upon bringing it home and placing it in the kitchen his wife (while preparing to bake the ham) pulled out the electric knife and cut one inch of both ends. The husband, observing this was curious.

> "The marvelous thing about a good question is that it shapes our identity as much by the asking as it does by the answering."
> David Whyte

"Why," he asked, "are you cutting the ends off the ham?"

"Because it makes it taste better." She replied, surprised at his question.

"Why do you think that?" He pressed, being polite and cordial, of course.

"Well," she replied pleasantly, "my mom taught me that."

So, the man goes to talk to his mother-in-law.

"Mom," he asks, "why do you cut the ends off the ham before

Lean Hybrid Project Delivery

you bake it?"

She thinks about this for a moment and replies, "because it makes it taste better."

"How did you learn of this secret?" the man queried.

"My mother taught me." She replied.

The man was very perplexed, it made no sense in his mind how cutting the ends off the ham would make it taste better. It just seemed to be a waste of good ham. So, he went to see the grandmother.

"Grandma," he began, "your daughter and granddaughter both cut an inch off both ends of the ham before they bake it. They say it makes it taste better. They tell me you taught them this. Can you explain how it makes it taste better?"

The old woman laughed and laughed.

"Sonny," she replied, "when grandpa and I were young and just starting out we were poor.

Smart:
- Specific
- Measurable
- Attainable
- Relevant
- Time-bound

George Doran, et al.

We couldn't afford a new stove and the hams would not fit in the oven so we cut the ends off so it would fit. We didn't want the kids growing up with 'poor thinking' so, I told them it made the ham taste better."

What are we doing that is "just the way it is done around here" that we should change or improve ... or maybe just stop doing?

We're closer to the Specific Measurable Goal by establishing the

Lean Hybrid Project Delivery

business environment and experience that is desired as the output of this project. We now know the goal in clear terms, we have developed a bulleted list of what we want and what we don't want as an outcome of the project, and we know what communications are required. This clearly defines the scope. Now we need to look very carefully at the time. When do we want this to be complete? This could be discussed at the beginning of the Conditions of Satisfaction meeting, but it sometimes helps to clarify the "what" before the "when". Sometimes, clearly understanding what you want influences when you want it. However, there is always a time element, now with a list of things that can be attributed to a successful outcome, we can review it based on the time limitations we have. This leads to another question: is the complete list attainable in the duration of time we have? If it is not, what are we going to remove? We now begin a process of deliberately prioritizing the Conditions of Satisfaction into a prioritized list of Conditions. This causes the team of executives and primary stakeholders to think critically about the clearly defined outcomes desired.

We have covered Scope and Time in the Conditions of Satisfaction, let's talk budget as well. As you know these have historically been classified as the Iron Triangle of Project Management. (Andrea Caccamese, 2012)

Lean Hybrid Project Delivery

In the case of most capital or strategic projects there will be document that is referred to (generally) as a funding document. This is the epitome of top-down budgeting. It is the authorization of the executive staff and controlling board to proceed with a project that includes the expenditure of millions if not billions of dollars as well as a high-level description of the

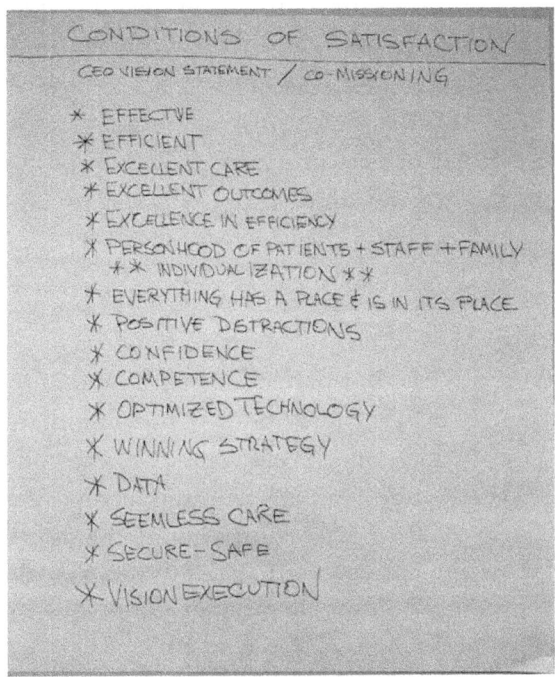

Conditions of Satisfaction sample

desired outcome. By the time you're in the process of building a Conditions of Satisfaction, you have generally understood what that funding document is working to establish and in the case of construction the schematic design and probably the design documents have all been published as well. So, to what extent are you able to influence the budget as it applies to the Conditions of Satisfaction? The obvious answer is that you are able to build within the confines of the executive orders what those Conditions of Satisfaction are, but you are also able to define more clearly what the desired outcomes were intended to be. In the Charette, which is the next phase of this methodology, you can begin to perform a bottom-up budget that is based on more accurate work breakdown structure estimates. In the

Lean Hybrid Project Delivery

Conditions of Satisfaction meeting, it should be discussed what the acceptable variance is between the top-down and the bottom-up budget assessments. Also, we should discuss the methodology for proposing budgetary changes relative to this project, as we move forward, based on what was discovered in the bottom-up estimate which is much more accurate than the top-down budget.

There are many conversations about whether frameworks should or should not be used in a modern project. The fact of the matter is that depending on how that project is formulated, there may be a framework. It is possible that if you are building a major new software development project you might have a framework that you're working within that provides you with a set of rules on what various component behaviors should be. This applies, particularly, to artificial intelligence projects, where one is building constructs of algorithms that can be moved back and forth between various different desired outcomes so as not to have to rebuild an algorithm for every single artificial intelligence instance. However, it also applies to other constructs as well such as construction constructs, that is,

> I characterize it as about a 70-30 to 80-20 rule, where the majority of your code has no idea what hardware it's running on.
> Tim Cowan, AE Software.

if there is an element of a facility that can be manufactured off-site and brought in (such as a hospital patient room headwall) there are rules that could apply that the Conditions of Satisfaction meeting should address at the highest levels and possibly at the level of directing that these should be investigated. There are things that we do that are not directly related to the core of the project, a medical headwall in a patient room is one example, but

Lean Hybrid Project Delivery

also in software environments there is the example of that 80% of the code should be able to run on any hardware platform, it's known as the 80/20 rule.

Emotional Intelligence
Finally, it is important for us all to understand that Emotional Intelligence plays a vital part in the success of any project. It is a pivotal element of success in the project delivery and therefore, critical for the Conditions of Satisfaction to address the impact and knowledge of the role of Emotional Intelligence. Fully 80% of the project success is built on Emotional Intelligence, while only 20% of the project success can be mapped to the mechanics of the project. This means that communication with and respect for the individuals who are performing the project has 80% more value than the mechanics the project. The steps that we will go through to deliver the project successfully only account for about 20% of the work that we'll put into the project.

Project success is about team and individual success.

Disciplined Agile Delivery
Disciplined Agile Delivery is the PMI's new poster child. The second edition of Introduction to Disciplined Agile Delivery was published in 2020, in the introduction there is a reference to the first edition being published in 2015. (Mark Lines, 2020) No one I have talked to remembers anything coming out of the PMI on the subject which is too bad. This is the most interesting publication with the PMI logo on it that I

> "Anyone can be angry – that is easy. But to be angry with the right person, to the right degree, at the right time, for the right purpose, and in the right way – that is not easy."
> – Aristotle

Lean Hybrid Project Delivery

have had the privilege of reading. That is a lot of verbiage to introduce a concept from this book that is applicable to this chapter. Chapter three of Introduction to Disciplined Agile Delivery is a nutshell view of the approach. I found it fascinating because it outlines some of the same tools and methodologies the book you are reading will cover.

One topic is new. They refer to this as the Exploratory Life Cycle. The idea is to minimize preliminary investments by performing small experiments to explore the potential of the perceived outcome. The output of these experiments is potentially a minimum viable product (MVP).

We may want to employ a version of this to explore scope, and as is always the case with the PMI … on page 19 they have published an "explore scope" diagram with ten high level processes to develop forty-seven knowledge areas. To see this diagram, which I recommend, go to PMI.org you will find the URL in the references at the end of the book. (PMI, 2021) This is not necessary to the Lean Hybrid Project Delivery methodology in this book. However, one of the key take-aways of this methodology is continuous improvement and continuous learning. This book would be incomplete without including it.

Clearly Defined, Specific, Measurable, Attainable, Relevant, and Time-bound

This will give us the capability in a single meeting to clearly describe the desired outcome of the project. What is the definition of the final milestone?

Having this clearly defined as we "Lean In" will help build teams, trust, commitments, and deliver outcomes. Using this guide (The Conditions of Satisfaction) we can better establish the flow of work and better eliminate waste. To that end, all meetings end with a plus/delta analysis.

Lean Hybrid Project Delivery

In this meeting, what did we do well? What can we improve in our next meeting?

Lean Hybrid Project Delivery

4 - Charette

Applied Imagination

Charrette
The term evolved from a pre-1900 exercise at the Ecole de Beaux Arts in France. Today it refers to creative process akin to visual brainstorming that is used by design professionals to develop solutions to a design problem within a limited timeframe.

"Coming together is a beginning. Keeping together is progress. Working together is success."
Henry Ford

This is a core meeting or primary tool in your toolkit. It is best done face to face with a Kanban board available. Best practice will include participants standing up "at the board". Each discipline should be given a different color "sticky" note. It is recommended to order 4" X 4" sticky notes of as many colors as you can. [**Pro Tip** – in the many Charettes I have coached … there has never been enough different colors.] Assure you have a couple of bright colored packs, you will need these for KNOWN constraints such as conflicting projects, holidays and any other hard dates, I like to include school start dates, spring breaks, and office party days. The milestones are shown at the top of the Kanban board by turning the sticky note 45 degrees forming a diamond. I like to use bright colors for milestones, but the selection is up to you. However, whatever color you choose try not to use that color for any tasks or activities.

You will need a couple boxes of black "Sharpy" pens so the team can write on the sticky note, and it can be read by the rest of the team. It's also

Lean Hybrid Project Delivery

a good idea to have a role of white paper (I've seen brown paper used as well.) and a couple roles of blue painter's tape to tape the white paper to the wall. Sticky notes can lose their stick, so it is a good idea to have a couple roles of cellophane tape available. (Both 3M and Post-It brands make super sticky pads which will save effort and they stick much better.) Use the same pattern of application even if you are unable to get square sticky notes.

This is a focused event, but it can also be fun. It's a great idea to bring refreshments and snacks, even lunch. Depending on the project priority and the corporate culture that may limit your time box, this meeting should be scheduled for four hours, or a group of one-hour meetings.

The early stages of Charette

We will approach the team with the attitude that we are working to learn what is needed from each other. Respect for the people we are

Lean Hybrid Project Delivery

working with is the corner stone of success. Focus on their definitions of value as it applies to the project delivery. What is important to them? Remember, being both the Project Manager and a technical expert or Subject Matter Expert can work against you because your team may 'let' you handle the stuff you appear to be a master of. However, we should also become increasingly proficient

> Semper Fidelis. Always Faithful.
> Semper ad Meliora. Always toward better things.

in the tasks we have to perform, people listen to us more seriously if we are the experts. (Portny, 2013)

Next, focus on driving out waste. We are still managing a project, we are not there for a group complaint session, we are there to collaborate. We are collaborating to deliver the project on time, under budget, with the resources provided. Collaborating is the single best way to drive out waste if everyone on the team is striving to do so. Focus on eliminating steps that are duplicated or are unnecessary.

Emphasize eliminating wait times and on working parallel efforts. Place attention on work happening in the team member's respective specialty that can be leveraged to get this project delivered. (Ballard I. D., 2016)

Continuous improvement of processes is a key element of this adaptation of Lean Hybrid Project Delivery. Continually ask, "How can we do this better, faster, more cost effectively, and with less work effort?" Semper Melius.

The group, the team, grows always more faithful and trustworthy to

Lean Hybrid Project Delivery

accomplish the goals of the project because they are working together as a team each knowing all the challenges each of the members of the team faces. We as Integrated Project Delivery Professionals now enable the team members to succeed.

At the beginning. As soon as we are assigned the project, we should begin asking with whom we collaborate. There is a problem-solving methodology known as "the five whys" – practice this methodology when selecting the "who" to collaborate with. Who asked for the project? Why? Was there someone else asking them? Who? Why? Who will be affected? What technologies will be employed? (Ohno, 1988)

In short, when we begin a project, we **begin** with collaboration because we are not the ones who are going to use the outcome, product, or service that results from our project. We are not the subject matter experts, we are not the medical professionals, and we are not doing the actual work to deliver the project. We are Lean Hybrid Project Delivery Professionals.

Engage Deeply
Collaborate – REALLY COLLABORATE with the objective of truly, constantly, and incrementally improving the process and delivery of projects. This means deeply comprehending the flow of work. It is sometimes difficult, particularly in highly siloed business cultures to breakdown the barriers to inter-discipline collaboration. In many cases, without executive participation, this will be your most difficult task. This is where your influential leadership skills will really come in handy. Don't worry if your influential skills need development. In Lean we are always improving.

Most Lean process promoting writers tend toward the Japanese words because so much is based on the work done by Eiji Toyoda through the years.

Lean Hybrid Project Delivery

"Kai Zen" which mean change and good. Together "kaizen" is the term used frequently to refer to the process of finding and making small incremental improvements in pursuit of perfection.

As I said in the introduction, I like to use Latin because it so echoes back to all those great military statements of courage, valor, and commitment, in the case of continuous improvement, Latin tends toward "semper melius." I like it better, but most Lean thinkers prefer to honor the Japanese words. However, I too will likely use Kaizen more frequently because it is easier to remember.

What is it about "semper melius" that applies itself to this topic?

A Charette in progress

Collaboration, should include all the stakeholders of the project, is a systematic study in pursuit of the perfect delivery of the project's product or outcome.

What does this mean in the context of the Charette?
The Charette is a brainstorming session. (Osborn, 1953) We are collaboratively imagining a decomposition of The Conditions of Satisfaction. This meeting should include the Vice Presidents (non-C-Level), and Assistant Vice Presidents, Senior Directors, and Directors. The outcome of this meeting is all the project milestones

Lean Hybrid Project Delivery

we will meet in delivering the Conditions of Satisfaction. These are identified by the business leaders to establish their priorities and requirements. We negotiate the milestone priorities wherever applicable; most milestones are sequential or time driven, those that are not, may be rearranged. Additionally, we will be looking at the FLOW of the project with a focus on Lean Principals. Where can the deliverables of the milestones overlap without adding additional risk to the project?

There will be a certain number of known activities or tasks that may be added to the project in this meeting (or series of meetings) depending on the size and complexity of the scope. These can be helpful in the next step, "the Flesh Out".

Your meeting should include the leaders of various departments including, but not limited to Information Technology (including telecommunications and the help desk), Marketing, Facilities Operations, and any vendors or contractors that will be directly involved in the successful delivery of the project.

Method: Milestone Planning
Our objective in this step is to build the Conditions of Satisfaction and identify the milestone that led up to those conditions being fulfilled. To do that we need a spectrum of team members ranging from the C-Level Officers and Senior Vice Presidents to those various department heads associated with corporate leadership who have any form of stakeholder role. (Kristen Hill, 2016)

This meeting should be held in a "Big Room". A Big Room is a conference room capable of supporting 25 to 30 people with available large whiteboards and space for pull planning. We will examine the physical processes of pull planning in a Big Room environment later.

From the executive level and that level that includes the senior technical or medical staff (if applicable), as well as from the user level

Lean Hybrid Project Delivery

that includes the appropriate customers, we are approaching the milestone effort as if we were filling out a "dream sheet". This means, we are looking for what they think is the best possible outcome of this desired project, as framed by the Conditions of Satisfaction. In their world, what is the problem to be solved, the enhancement to be installed, the workflow to be improved, or the work effort to be eliminated. What is the strategic product to be delivered?

We are pulling backward from this ideal set of Conditions of Satisfaction, in a virtual retrospective, walking backwards through time, along with the technical and business experts that will be managing the work required to accomplish the desired outcome. As we begin to perform this exercise, certain discoveries begin to be revealed. As an example, (assuming there is an expansion of offices or an new facility) it may be revealed that this outcome will require additional workstations, these workstations may need additional computing power as compared to the existing workstations; the existing workstations may need to be upgraded; additional network jacks may need to be installed, and additional network switches may need to be ordered to support the additional ports required. Servers may need to be ordered and time will need to be spent determining what is the optimal operating system for the servers. What additional support staff will be required, both in the desktop the server and the software specialties? These questions are translated into high-level activities or possibly milestones that are passed to the Flesh Out sessions (or meetings or conversations). A milestone may be (in this case) "IT Infrastructure Defined".

The purpose of this meeting is to identify milestones not to identify costs or difficulties. What milestones need to be identified to reach the optimal Conditions of Satisfaction set forward in the dream sheet.

Method: Establish ROM and GMP
In another iteration of the same conversation high-level Subject

Lean Hybrid Project Delivery

Matter Expert opinions may be discussed among the chief stakeholders and end users to perform "Guaranteed Maximum Price" and / or "Rough Order of Magnitude" estimates and reach a determination of what is the maximum budgetary limit the owner is willing to spend to reach the optimal Conditions of Satisfaction. In this iteration it may be determined that if the cost exceeds a certain level the Conditions of Satisfaction would be altered or diminished to provide a working model of a product that is satisfactory to the stakeholders and end-users that may not reach the highest level of their original desire.

Depending on the size of the project this meeting may be required multiple times; but it is important that this meeting occurs and that you have a plan of what the final milestone and all the milestones leading up to it are. Each of the milestones should be separated by the discipline area or specialty that will be performing the work. These separations form lanes of responsibility are referred to as "specialty swim lanes", and we will discuss that later in this book. (Seed, 2017)

Method: Pull Planning
Thinking of this process as a micro pre-mortem is often helpful. Imaging the Conditions of Satisfaction as complete and then "pulling" back from the completed project thinking through the steps that were accomplished to successfully complete the project. This is called "Pull Planning". Pull Planning, in simplest terms, is a technique that is used as part of developing a coordinated plan for one phase of a project. The purpose of pull planning is to design a project-based production system in conformance with Lean principles. Pull planning is a collaborative process that can help to get buy-in from all project participants using a backward pass, "pulling activities into the process."

Lean Hybrid Project Delivery

Pull planning is a lean process that replaces many predictive scheduling techniques and works by identifying activities that are compressed to reduce overall activity duration.

The "pure" Lean or original concept of a pull system is a Lean technique for reducing the waste of any production process. Applying a pull system allows you to start new work only when there is customer demand for it. This allows you to reduce overhead and optimize storage costs. It is important to understand this (even though it applies in its original form to manufacturing) because we want to know who our customer is and what their expectation is.

The milestones that are the output of the Charette should include thought as to what can be delivered as a working component of the project at the earliest opportunity. While this is difficult in construction because it is difficult to turn over a new building until it is complete, it does apply to the various trades and subcontractor activities that tend to be thought of as sequential.

In KANBAN Planning "Pull Planning is "pulling" out of the Backlog into the "Work In Progress" column.

Though there are three contexts to the term "pull" planning as it applies to the Charette, we are pulling back, performing a backward pass from an imagined finished project. We'll cover more on this in the Flesh-Out and Sequencing chapters of this book.

In predictive project management one of the most difficult problems is the perceived inability of humans to predictively plan the end from the beginning. The planning fallacy refers to a prediction phenomenon, where people historically underestimate the time it

Lean Hybrid Project Delivery

takes to complete a future task or group of tasks that compose a project, despite knowledge that previous tasks have generally taken longer than planned. The pull planning methodology helps to overcome this issue by focusing the team on a collaborative 'backward' pass through the project flow of work.

Method: Current State Review

Finally, it the Charette, it is worthwhile to discuss the Current State. Remember the Lean analogy of Grandma's Ham in Chapter Three, we want to assure that we expressly leave out what we do not wan in the project. We do, however, want to assure we keep what we want as well. We'll evaluate a future state in more depth as we move through the book.

Method: Identify Known Risks

This team of individuals is the most likely group to have a grasp on business impacting risks. As you are pulling back from the final miles stone (defined by the Conditions of Satisfaction) be thinking about the risks that may be encountered. These can be fleshed out in more detail as we get into the next phase of planning with the line managers.

Lean is about flow of work and elimination of waste, while always improving. All meetings end with a plus/delta analysis.

Lean Hybrid Project Delivery

5 – Flesh-out

> ### Team Building Part 1
>
> **Team Building**
> The Flesh Out is an intensive, interactive, ideally face to face activity. It is the first to include Directors and Line Managers. The Lean Construction Institute refers to the Line Managers as "Last Planners". This works well with other types of planners as well. Including the managers who are responsible to get the work done builds commitment and trust and establishes the team.
>
> "Good teams become great teams when members TRUST each other enough to surrender the ME for the WE."
> Phil Jackson

From many one - *E Pluribus Unum.*

Respect for people. Its basic really, and it is the basic building block for this methodology. Consistent with both Agile and Lean principles our goal is to develop a culture of respect and continuous improvement. (LCI, 2017) (Agile Alliance, 2017) As Integrated Project Delivery Professionals we strive to build respect for people based on team building and commitment stability. We can depend on each other to make and keep commitments because we are making these commitments to each other and continuously working to improve our ability to deliver on the commitments we make. It is important to us to learn as we grow. Initially our commitments are less accurate than they are after a cycle of iterative processes of planning and committing. The more iterations we cycle through the more accurate our plans become. As we develop our skills for planning, committing, and delivering we become more confident, not only with ourselves, but also with the other members of our team.

Lean Hybrid Project Delivery

This may take several meetings.

Step by step - *Gradus per gradus.* This is one aspect of how we develop trust. As each of us commits to continuously improve our delivery commitments and accept the delivery of those commitments from the team members we have received commitments from – and the consistency of the deliveries increases – trust grows within the team. Through this process of constant improvement (semper melius or kaizen) team members begin to feel increased pride in their own work positions as well as finding satisfaction with the team. We desire that every member of the team feels they are as important as every other member and that the team is the most important thing in the project process second only to the delivery of the project which is the purpose of the team. We are consciously working to reach the best possible resolution of the top three levels of Maslow's Hierarchy of Needs. (Fowler, 2014)

> "If everyone is moving forward together, then success takes care of itself."
>
> Henry Ford

This is aimed at creating more value for our customers and stakeholders through Lean Hybrid Project Delivery. The Lean Hybrid Project Delivery System is an organized implementation of Lean Principles and Tools. Together the team can employ these to function in unison to deliver the project. (NASA, 2017)

Lean Hybrid Project Delivery

Toward perfection – *Ad Perfectum*. Humans are notoriously poor planners. Our forward vision is great as it applies to a goal or ultimate objective, such as built buildings or climbed mountains, but the process of planning all the individual steps necessary to actually build the building or climb the mountain are seldom accurate. No matter how many networks we have built, buildings we have built, or mountains we have climbed, each network, building, and mountain is different. In Healthcare Information Technology every project is somewhat different from every other one because the stakeholders are different, the desired outcomes are different and without fail the function we as project managers are supporting is literally life and death. The outcomes we desire in Information Technology are accurate, usable, data that can inform knowledgeable decisions by clinicians and researchers to heal and prevent disease, wounds, and damage. In our case as Lean Hybrid Project Delivery Professionals this includes a complex weave of infrastructure, software, hardware, interfaces, and methodologies. It is impossible for a single person to understand all the various aspects and complexities of the system. To become more nearly perfect with each iteration requires that

> "The first step is transformation of the individual. This transformation is discontinuous. It comes from understanding the system of profound knowledge. The individual, transformed, will perceive new meaning in his life, to events, to numbers, to interactions between people."
>
> W. Edwards Deming

Lean Hybrid Project Delivery

we collaboratively plan. To do this we include all team members to the level of the last planner. The Last Planner® can be the last person in the chain of command that has the responsibility of planning the implementation of the product of our project. This can be the Team Lead in the Network Infrastructure group or the Nurse Manager. (Glenn Ballard, 2016) (Ortiz, 2017)

1. Make leadership, including decision-making, a shared activity within the team.
2. Make sure that you're holding not just the individuals accountable, but the team accountable.
3. Ensure the team understands its bigger purpose for existing.
4. Empower and coach team members to be problem solvers themselves.
5. Measure the success of the team's performance, not just each individual's performance.
6. Foster communication and healthy conflict between team members.
7. Work to build empathy and trust between team members.

Method: Pull Planning - *Trahere concilio*. Pull planning has two primary origins. The most commonly understood seems to be the idea of a 'virtual project retrospective'. The term retrospective is often used in a post-mortem thoughtprocess. In this methodology, after the project is complete, the team gets together and reviews all the things that happened both good and bad to determine what could have been improved upon, what went well, and what failed and must be reinvented before the team attempts another project. A virtual project retrospective could be referred to as a 'pre-mortem'. Imagining the project is

Lean Hybrid Project Delivery

complete or the milestone is reached and then imagining all those things as if they had already happened, pulling back from the completed project and decomposing the activities and actions that were required to accomplish that successful completion.

The author coaching pull planning

The other, and the one I like better, is the concept of pulling a part from the shelf in an auto parts facility in a car dealership.

The part can be anything really, from a can of paint in a warehouse home improvement store to a pack of rice in a grocery store. The auto part is best in my mind as an example. Let's examine a simple part and pull-plan backwards. (Womack J. , 2003)

Let's use a headlight, simple enough, right? So, imagine you are standing in front of the shelf. The headlight is exactly where it should be, in the packaging it should be in, properly labeled, priced, and available. This is the state the product needs to be in to be usable by you as you fill customer orders. The milestone is

Lean Hybrid Project Delivery

'parts in stock for sale'. It is the last step of the process that allows you to open your parts store to the customer.

What had to occur for the packaging to be correctly printed, shaped, cut and formed? Thinking backwards from the finished product, what was the last function to occur? You are correct if you suggest that was placing the light in it. Prior to the light placement, the packaging was folded into a box.

Who or what folded the box? Who or what designed the cut of the box? Who designed the print format of the box, wordsmithed the verbiage on the box, and added the logos? Was the box printed in a flat format and cut? Remember you're thinking backwards from the finished product. In either case, this process is framed by the questions: "What was the last thing done to make this milestone complete?

Who did this? How long did it take?" We're going to spend the next few chapters developing this methodology, but it is important now for us to start thinking who do we need to collaboratively plan with in order to reach the milestone of 'usable product'?

Lean Hybrid Project Delivery

Method: Chunking

As we start decomposing the final milestone, as defined by the Conditions of Satisfaction we find the participants have no frame of reference as to what to write on their sticky note. The concept is initially foreign to us all. We are forcing ourselves to think in reverse … to pull plan back from the finished outcome.

Using the example of the headlight; one might come to a decomposed activity of removing the packaged headlight from the shipping container and placing the packaged headlight on the shelf. This is great but keep chunking it down. How do we know where the appropriate shelf location is? Is there a reason the location is where it is? Who decided that it is in that specific location? Did someone label the specific location? Do we know what the appropriate stocking levels are? Who decided that?

In many cases as we begin the process of decomposition, we are confronted with coaching the team into a deeply collaborative thought process they have never experienced before. As we begin this exercise, we also begin to discover that we 'build in' lead times and lag times based on past experiences. (Lindley, 1966)

> Chunking down is getting more detail by probing for more information about the high-level information you already have. The goal is to find out more, fill in the empty gaps in your picture, test the reality of the situation, and so on. The more you ask chunking questions, the more you will find further detail.
> ChangingMinds.org

Lean Hybrid Project Delivery

Our objective is to chunk the activities down to a level where each activity can be completed by an individual or a small group in five days or less. Chunking down has its origins in Emotional Intelligence. It originated as a tool to help people cope with large problems that caused them emotional stress. By chunking down the problem to smaller pieces it helps get your mind around the pieces easier. We are doing the same thing here. By chunking down the "problem" into smaller individual tasks it makes it easier to see exactly what activities need to be accomplished to deliver the project.

In this step (the Flesh Out) we are working collaboratively to fill in all the blanks as to what necessarily needs to be done. However, we are not necessarily addressing WHEN they need to be done or IN WHAT Order they need to be done. The process will by its own nature sequence the events in a near correct order, but we will revisit the sequencing in the next step in the project.

Don't sweat the sequence in the decomposition meeting. Our objective here is to flesh out the activities into their component parts and to strip off built in lead and lag times that are intuitive to the planning team.

One of the outcomes of this work effort is a pretty accurate work effort estimate. A bottom-up estimate of person hours it will require to accomplish the completion of the Conditions of Satisfaction.

Lean Hybrid Project Delivery

Method: Making Commitments

In the image below we see an example of a sticky note that was used in another of my projects. While you can develop any method you feel comfortable with, this one works well so I recommend it. In the upper left corner is the name of the individual that will be completing the activity. In the upper right corner is the duration of the activity. As you can see this one exceeds the goal of five days. In a later effort we decomposed this activity further by segmenting the installation effort into zones in the facility. In the center of the upper row is a number – well come back to that in the Sequencing chapter.

In the center row is the activity to be completed. In the lower row is the prerequisite for that activity to be completed. There is another sticky on the Kanban board that has the "Sonitor Equipment Delivery" in the center as its activity and an individual who will be delivering the equipment. Both Jim R. and Tirone B. are in the room together collaborating on the FLOW of work. They are committing to each other. This may be occurring with Jim R.'s manager in the room assigning this activity to Jim, and the sequencing may be done by Jim directly, but we will come back to that later.

Pull Plan Sticky Note Example

Method:

Lean Hybrid Project Delivery

Precision vs. Accuracy

It's worthwhile here to talk a little bit about the difference between precision and accuracy. One of the things that we are working on as we're working on building commitments and trust among the team members is accurate predictions or accurate estimates on the work effort that needs to be accomplished. (Tan, 2020) This means that we are looking to find a way to assure that when a team member says they can do something in five days or less that it means they can do something in five days or less. The way to do that is iteratively. As each of the individuals is estimating that it will require five days for them to complete the item that's on the task (once the prerequisite has been met). Then on a weekly basis, when we meet, we ask, "Have you completed that task?" By doing this, we are fine tuning our ability to estimate. None of us is going to be good at estimating at first. We have no frame of reference other than experience. If we're doing something that none of us has experienced before then we are estimating based on a hypothesis that is also based on experience or education. Therefore, in the first few iterations of planning (and marking planned activities complete if they are complete), we are maturing in our capability to be able to estimate properly. As individuals get more comfortable with this, we will find that they are better able to live up to the estimates that they give us.

> Why estimate? To get a clear view of the project *reality* so you can make informed decisions to control your project to hit your targets.
> — Philip Tan, MIT

In the illustration of the targets that follows, you will see that in

Lean Hybrid Project Delivery

target both targets there are four holes (assumed to have been put there by bullets for instance). Which is more precise, and which is more accurate? The holes in target B are closely grouped and are low and to the left. It's a precise grouping. In target A you will notice that none of the holes are in the bullseye either. However, they are all around the bullseye of the target. Here we have some accuracy in hitting the target. In estimating duration or cost, accuracy is more important to us than precision. However, precision is important. As we begin to understand how the estimating members of the team are performing their estimates, we might find that someone is precisely off by a couple of time periods but precisely estimating incorrectly every time. This is good because it enables us to help that individual move their target groupings up into the right, so that they can be on target every time. Similarly, for the one who is more accurate, that is the one who is accurately estimating all around the target but never actually on target, we can help them also to be more precise and more accurate and move their group closer to being on target at all times.

This will be continuously improved and continuously adjusted as we go through the Sequencing part of this methodology in the next chapter. Although a great deal of the chronology of the activities will, by the nature of pull planning, be in the correct sequence at the outcome the Flesh Out segment of the planning we continue to have the existence of the Sequencing segment because we are moving ever closer to the individuals who are actually doing the work. The individuals who are actually accomplishing the work or their line managers are the ones who are most accurate in breaking the activities down into the actual

Lean Hybrid Project Delivery

work that must be done.

So, what are we saying? We are saying that while the executives gave us a clear description of the outcome that they were looking for in the Conditions of Satisfaction – and – while the upper level managers, senior vice presidents, vice presidents, and senior

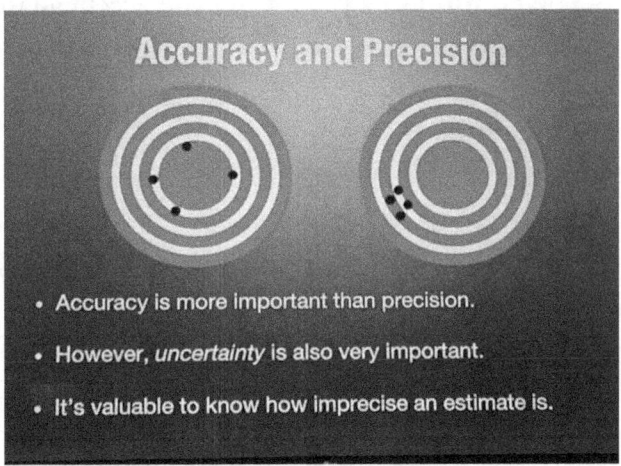

From MIT OCW CMS608

directors gave us milestones it the Charette - and – while the directors, upper managers, vice presidents, other mid-level leaders here in the Flesh Out are filling in the gaps of the activities and processes that need to be in place to accomplish the milestones - we are still adding milestones and higher level activities. However, in the Flesh Out we are trying to break them down to five-day increments that can be completed by an individual or small group. We are doing this for the sake of accuracy in our estimations, but it isn't until we get into the sequencing that we can really feel confidence that everyone including the people getting the work done are in agreement.

Lean Hybrid Project Delivery

Method: Estimating vs. Planning
- You **plan** to achieve a desired target.
- You **estimate** to see if the plan is realistic.
- You **commit to an estimate** when you think it is accurate.
- You **commit to a plan** when you accept a task.
- You **re-estimate** as you work on the task.

Tool: Estimate vs. Plan
- "It has to be done in 2 days" is a **Target**
- "It'll take about 2 days to do it" is an **Estimate**
- "I'll need 2 days to do it" is a **Plan** *(Signifies ownership)*
- "It took 2 days to do it" is **Reality**

Method: Applied Lean

As we're doing this, it is important that we are constantly reminding everyone who's participating (and remember ourselves) this is a highly collaborative effort. We have many people participating in this planning operation (from mid-level management and above) working through fleshing out what is in the milestones and what are the activities that need to be completed. Among the things that we are constantly looking for is:

- What are the activities that do not need to be done?
- What is the Muda?
- Where are we employing activities or rework or waste that is not required to actually deliver the product that is defined by the conditions of satisfaction?

This is important as we move through the Flesh Out because we

Lean Hybrid Project Delivery

are working diligently and collaboratively with everyone to streamline the workflow to assure that we're not working any harder ... we're performing any more work than is actually required to accomplish the task.

One of the outcomes for every activity task on the sticky notes is an activity owner assignment. Who is actually going to perform the work? That individual should take ownership in participation with the team not as a conscript.

Tool: Current State Validation
Building on the Current State Review performed in the Charette, we will include in this effort a review and validation of the current state. While a management review is accurate, it is important to fully understand the impacts of some current states. For example, at this level of validation we are not only looking at the fact that there is a work around, but also what is the impact of the work around. This may be more important in a project that is altering something such as a renovation or an update to a major software program than it would be to new construction, or a new application being built from the ground up. However, even in the current state of "null" there is a current state consciousness of the individuals who are involved in the project of things that have triggered the application or the construction. None of us is approaching these things from a clean slate, we are approaching them from a position of experience and education. It is important that we review the current state and clearly understand how that state reflects on the situation that we are trying to develop. Current state review is an integral part of the Flesh Out as we determine what to leave in and what to leave out based on that exact segment of the Conditions of Satisfaction. Therefore, the current state review, even if it is totally new, gives the

Lean Hybrid Project Delivery

participants of this collaborative planning session the ability to clearly understand what the outcome they desire is based on either of their education and experience or an existing current state.

Tool: Future State Review

The future state review is a tool that enables us to 'look' into a finished product and 'see' it in its operational business environment. As an example, in a recent project where I was involved in the construction of a three-story vertical expansion that provided state of the art ICU rooms in a hospital environment, I asked the leaders in the room to close their eyes and walk with me down the hall. I see rooms on their right and the nurse station and support rooms on their left.

- What are they seeing?
- What does it smell like?
- What does the floor look like?
- Is the air conditioning hot or cold?
- What are the sounds that the monitors are making?
- Do the nurses know where everything is?

This has value because the visualization of the future state sometimes triggers things among the individuals who are performing the future state review enabling them to catch things that they were overlooking. This applies to a software application as well because sometimes viewing it from the perspective of the user, or the customer, in its future state enables us to see the customer happily using it.

- what are they doing?
- what is their experience?
- and what is their expectation?

Lean Hybrid Project Delivery

Also, there's the objective of a change. If you're doing a major change in your project, you have an old way of doing your business environment (user experience) in your application and there will be a new way of doing your operation in a business environment that has been changed.

The future state review is begun here, not in the Charette because we are building on the Conditions of Satisfaction and the Charette to define a plan for a future state. An output of the Sequencing chapter is a Future State Validation which includes the leadership, key stakeholders, and executives (if appropriate). Depending on the team dynamics, the executives may reserve a right of approval of the future state, as may any of (or all of) the primary stakeholders.

Case Study
Years ago, I was involved at a Fortune 500 transportation company. We were doing a software upgrade to an important part of the business operations. We had extensive meetings where the Current State Review and the Future State Review so that the customers, or the users, who were also employees of the company would understand how things would be different as they were working on the computer screen while performing their jobs. After extensive visitation on this, it was determined that we needed to alter the future state user interface to more nearly model the current state user interface. As it turned out in this particular project after we had done the upgrade and done the extensive amount of work ... when we implemented the updated software ... the users came back requesting changes that eventually ended up very nearly what the future state would have been had we not made changes at all.

Lean Hybrid Project Delivery

I include this story because at the time we were not doing Lean Hybrid Project Delivery methodologies. In this methodology, we would have been collaborating with the stakeholders and validating outcomes all along the process, concentrating on Business Environment from the beginning until the end. In this methodology we are delivering a project not closing it.

Review Risks

Reviewing risks is an ongoing, iterative process. In the Charette the team identified known risks. Here we are reviewing and evaluating those risks, creating a list of risks with potential responses, and assuring there is sufficient contingency for the known risks. The outcome of this review is a list of known risks with 'Jeopardy Codes' assigned. Jeopardy Codes are agreed to by the team. They can be assigned by straight number (i.e., 1-10) or through a Fibonacci sequence (i.e., 0, 1, 1, 2, 3, 5, 8, 13, 21, 34) if the risks have a compounding impact on the project.

It is important to remember the positive risks when performing this team exercise.

Review Alternate Plans

This is an iterative process and will be reviewed in project meetings throughout the delivery of the project. Part of being successful in a project is having alternatives. Here we review the risks and plan for alternative methods to accomplish the same or a similar acceptable outcome.

Change Management Process

This is an iterative process and will be maintained throughout the project. Here in the Flesh Out we will assure the change process is completely understood. If there are corporate policies

Lean Hybrid Project Delivery

governing changes, we will include them as part of the project. If there are not corporate change policies, we will collaboratively produce change policies that properly inform the team. This indicates we will publish who on the team from the key stakeholders to the line managers require information, involvement, and has authorization responsibility for each change. Also, we will classify changes. Changes that impact the finances of the project will likely be handled differently the minor schedule changes.

Develop a Test Strategy
This particularly applies to Information Technology, Finance, and other business system projects but also to manufacturing, implementation and integration projects. Time should be spent to discuss, plan, and specify a test strategy which includes:

- What should be tested,
- When should it be tested,
- How should it be tested,
- and what are the specific acceptable results?

Lean teaches us to test early and frequently. We should not wait until the end of a product run to test. Rework is waste. Since this book is designed to cover projects globally, I encourage you to read up on this subject as it applies to your project. Simply stated though, test at the earliest possible time with the earliest Minimum Viable Product (MVP) and at the earliest point where integrations can be validated.

Develop an Integration Strategy
In the modern business environment very few things or groups of things being affected by a project stand alone. Everything is part

Lean Hybrid Project Delivery

of a system. (Sloan, 2021) Apply Little's law. Often referred to as the First Law of Operations, Little's Law is named for John D.C. Little, a professor post tenure at MIT Sloan and an MIT Institute Professor Emeritus. Little proved that the three most important properties of any system — throughput, lead time, and work-in-process — must obey the following simple relationship:

$$L = \lambda * T$$

(Work-in-Process)   (Throughput) *   (Lead Time)

It is important to remember that, as a part of this methodology, we are specifically looking at the Business Environment as it will be at the delivery of the project. We are performing **transition planning** from the beginning of the project to the end of the project. (Dimitris Bertsimas, 2001)

The integration strategy should include all the members of the various teams, from Information Technology, Marketing, Finance, operations, procurement, engineering, to the production floor management.

As Lean Integrated Project Delivery professionals, we are surrounded by systems. The products, processes, and projects that we work on are increasingly complex and interrelated systems. Organizations are calling on you, their project professionals, to drive and optimize complex projects under high-pressure conditions. With this methodology, we do that collaboratively, incorporating the team that is talented, capable,

Lean Hybrid Project Delivery

and confident to complete the associated aspects, planning, design, and construction of the activities that successfully develop successful project deliveries.

Now is the time - *Nunc est tempus.* Here we are reading this thinking when do we get this collaborative team together?

When is our project supposed to start? When is it supposed to be complete? In most environments we work in, the best answer is something like "yesterday"!

> "The people who make the algorithms do not always know where the value in the facility is,"
> Sergio Garza, VP Heineken Mexico

As frustrating as that is, it is important that every day that passes without the finished product of the project fully supporting the business environment is another day when the leadership is making decisions based on incomplete data and insufficient knowledge. This is not saying the leadership is not knowledgeable, it is saying we are all working daily to get better, to learn more, to be more accurate in our assessments and decisions. Data gathered is the seed that germinates into information as it is stored properly in databases that are usable, but the fruit of that information technology is knowledge.

Lean is about flow of work and elimination of waste, while always improving. All meetings end with a plus/delta analysis.

Now is the time.

Lean Hybrid Project Delivery

6 - Sequencing

Team Building Part 2

Team Building
The Sequencing planning effort is an intensive, interactive, ideally face to face activity. It is the next to include Directors and Line Managers. The Lean Construction Institute refers to the Line Managers as "Last Planners". This works well with other types of planners as well. Including the managers who are responsible to get the work done builds commitment and trust and establishes the team.

"Individual commitment to a group effort--that is what makes a team work, a company work, a society work, a civilization work."
Vince Lombardi

This is an iterative meeting. It may take more than one meeting to accomplish the first "final" plan. However, it is likely that you and the team will be revisiting this weekly or by weekly or monthly, depending on what the team determines is an appropriate level of analysis.

Method: Focus on True North
Imagine you are flying an aircraft, even if all you are doing is practicing or site seeing, you must by necessity know where you intend to land. For the purposes of this book imagine that you are flying from place "A" northward to place "B" in a straight line. As you become airborne you remember for the entire trip you will have a crosswind from one side. You are also aware the crosswind is gusty rather than steady, so you will be unable to maintain a simple adjustment but will need to adjust frequently to varying amounts or you will be blown off course. If you were only traveling a few hundred feet this would be no problem, you would be able to see your target runway and adjust to it rapidly. However, for the purposes of this example you are traveling for hundreds of miles and the wind intensity is sometimes severe.

If you do not keep track of your

Lean Hybrid Project Delivery

position relative to your planned course and adjust back to it, you will soon be lost and have no idea where your intended landing zone is.

Projects are often similar to this example. We start off with an understanding of what we are managing to but along the way there are winds of change blowing us about and, in many cases, we have no way of pushing the project back on course. Instead, we spend hours and hours attempting to salvage the intended goal and just somehow get the project online or live and get out of it with our sanity.

The concept of "true north" (Ammer, 2003) is an American idiom that originated in the late 1900's that refers to the concept that we must have an absolute purpose and focus toward which we are steering. Without this, we are subject to meandering off the path and becoming lost in the "Muda" (waste) of lost time and effort. (Lean Enterprise Institute, 2017) Worse yet would be to become so off course that we are lost in the waste land – *vastum terram*.

The number one reason for project failure in is scope creep and uncontrolled change. It is notably one of the top three reasons all projects fail. (Project Management Institute, 2017)

If you don't know where you want to go, then it doesn't matter which path you take. So said the Cheshire Cat in Lewis Carroll's classic book. (Dodson, 1865) This is an absolute in the world of Lean Hybrid Project Delivery. Defining **true north** at the outset of the project in the Conditions of Satisfaction gives us a much better chance of delivering the project on time, on schedule, and on Budget. Otherwise, every time we come to a fork in the road, we are taking a chance our choice may lead to a failure to deliver the project on budget or on time or at all. This is why we did the Conditions of Satisfaction process.

True North is not a new concept, it has been taught since earliest Greek Olympics, when the concept of laying off all the things that

Lean Hybrid Project Delivery

weighed a runner down led to their competing naked. Similarly, sculptors cut away anything that doesn't look like their subject.

In more recent times, Toyota implemented the idea that all waste in the production system should be cut away. The basis being the least amount of effort needed to produce a quality vehicle was the best method of manufacturing. (Ohno, 1988)

Up until now all the chapters in this book have been leading us to this point. This particular "how" has been founded by all the things written in this book so far. Now we are challenged as leaders and coaches to bring the integration of teams into alignment helping to coach and guide them to a lean project delivery that includes their various views and desires. By following this Lean Hybrid Project Delivery methodology, we will learn together as we progress in our profession.

Method: Big Room Sequence Pull Planning
Use the Big Room concept and Pull Planning to apply the principles of this system to develop a properly detailed and sequenced plan. This part of the planning includes the Directors and Line Managers. These are the stakeholders closest to the work. These are they who will lead the actual work effort, collaborate with parallel teams, and manage risks. These are the owners of the activities. They know better than anyone what exactly it will take to get the work completed and what the prerequisites are.

"This is an explicate plan formed by the end users of actual requirements that must be satisfied in order that they may feel that they have received exactly what they wanted." (Kristen Hill, 2016)

Collaborate deeply with all the parties. How often have we started a project thinking we understand what the end user or chief stakeholder wanted but once we got involved, we learned there was

Lean Hybrid Project Delivery

some major element missing for the implementation that was going to require a lot of time and a lot of effort to resolve? These "last planners" as they are referred to by the Lean Construction Institute are those who know the work effort.

Tool: Building a Team

This is an appropriate time to tell the story that Patrick Winston, MIT Professor of Artificial Intelligence, in his How to Speak course. (Winston, 2018) In this course he tells of being at a large ceremony in which the visitors or participants were sitting at round tables of approximately 10 people. He was seated next to Julia Childs. Throughout the process of the ceremony people were coming up with napkins, books, and papers asking Julia for her signature. After a while Patrick leaned over and asked what it is like to be famous. To which she replied, "Oh, you get used to it." Patrick then, in his talk illuminates on his takeaway from that conversation. He says, "You get used to being famous, but you never get used to being ignored." Our takeaway from this story is - to

> You get used to being famous. You never get used to being ignored.
> — Patrick Winston, MIT

successfully manage a group of people in a Lean Hybrid Project Delivery methodology, it is the responsibility of the project management professional to make the talent famous. Find what they are doing well and promote it. Find the small successes and applaud them to the upper management. Create an atmosphere around the talent that makes them what to do their best. As a Lean Hybrid Project Delivery Professional, you are not here to become famous, you are here to be a servant leader promoting the fame of those in leadership and those in the ranks. The talent, as I like to call them, are the ones who deserve to have their names in the credits.

> The process of Sequencing is a process of individual activity owners reviewing their induvial activities. They are aligning tasks together by saying "I can do this, in this amount of time, if you can give me that." The outcome is a modified estimate that is more nearly accurate.

Tool: Tie in Prerequisites
In the Flesh Out we added many tasks in somewhat of a chronological order and to some extent we included prerequisite tasks that needed to be completed before a given task could proceed. We also added assigned ownership of those activities and tried to chunk them down to five-day periods or less. Now, we are meeting with those individuals who are either completing the activities or directly supervise those individuals or small teams that will be completing the activities.

Be prepared! This is one of the most difficult parts of this methodology. These are the individuals who know what to do however, they may be unaccustomed to participating in their own direction as to how to do it because they have always been TOLD what to do, in this case you must coach them into participation. Sometimes, you may find the talent is initially offended because they

Lean Hybrid Project Delivery

feel they are being questioned about their commitment or their competence. When people are new to the methodology, we need to guide them into understanding and trust.

We are asking them to review these activities that they may not have had a say in developing. As the Lean Hybrid Project Delivery professional, you are now challenged with incorporating their opinions, knowledge, training, and talents into the project. Use the "make them famous" approach.

At the end of this exercise, one should be able to answer the questions:
1. What needed to be done before I can do this activity or task?
2. What comes next, who is doing that activity or task?

Pro Tip! Be Real.
Be sincere in your praise and appreciation, if you do not sincerely appreciate the level of work, training, experience, specialty, and commitment these individuals are bringing to the project they will not appreciate you, nor will they accept your servant leadership.

Trust is built on commitment. Commitment is built on truth. Truth is built on transparency. Transparency is the result of servant leadership mentality that you are truly not superior to the team, and you live up to your commitments.

Tool: Identify Constraints
Using estimation modifiers, such as, "I can do this **if** I have this" systematically identify constraints. Constraints can be prerequisites, but let's endeavor to frame constraints as outside limitations on time such as lead times or regulatory delays. Constraints can also be training and education constraints. In the context of this project management methodology constraints are not necessarily prerequisites as an example - Thanksgiving and the day after

Lean Hybrid Project Delivery

Thanksgiving - are constraints on the workforce. National holidays are constraints on the workforce. Time zones, language, international legal and cultural parameters can be constraints. Available qualified and trained personnel is a constraint on the project. Other competing projects can be a constraint, particularly if the same talent or other resources are in use in the competing project. Therefore, for the purposes of this project methodology constraints are not prerequisites but are anything that limits the ability to work under normal conditions. In the context of Kanban, where we are putting sticky notes on the board to identify known constraints, these sticky notes should go at the top of the con bon board, on the same level as the milestones. You can see an example of this in a picture in this section, the small brown square below the sticky note boldly labeled WK5 is a holiday constraint.

Tool: Set Priorities

This has been happening intuitively as we did the flesh out and as we are doing the review of the constraints and prerequisites. However, it is worthwhile to call formal attention to priorities. As the team is reviewing the prerequisites and constraints it is worthwhile to periodically step back and ask, "are any of these activities and their subordinate activities higher priority than what we're showing them on the Kanban board?"

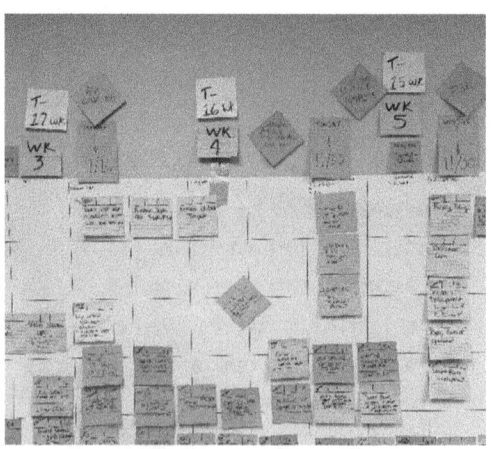

If they are, then we need to take this time to adjust their location in the flow of work. Remember in Lean it's all about the flow of work and the reduction of waste. Every time we have a project activity that is waiting for no cause other than it just hasn't been prioritized properly, we are

Lean Hybrid Project Delivery

wasting time and that is Muda that needs to be removed, eliminated, or reduced as much as possible. This is a key principle in the pull planning methodology, and it is incredibly important that this becomes a key in our thought processes as we pull plan this project to a more nearly perfect plan. The other key principle we are addressing is constant improvement.

Pro Tip: Lean In
As Lean Hybrid Project Delivery professionals we are always "Leaning Into" the project. This means we are continually looking for waste and Muda. At the beginning and until the team members begin to mature in their developing skills, we are always looking for the activity or task that has lead times, lag times, and wait times built in. These are normally added by individuals when estimating because in the old way of doing things one always had to wait for someone else to deliver before proceeding. In this colaborative model we are working together to determine what is real, so the lead time from the previous activity should not be built into the current one.

> Of all the things that can boost inner work life, the most important is making progress in meaningful work.
>
> Teresa Amabile, HBR

Celebrate work not done!
(Teresa M. Amabile, 2011) When we think about progress, we often think about how good it feels to achieve a long-term goal or make a major breakthrough. These big wins are great but they usually rare. The good news is that even small wins can boost inner work life tremendously. We shoud be working with the team members to find those small, incremental, and sometimes overlooked efficiencies and celebrating them as we find them. In this way the team members are encouraged to learn to "Lean In" as well. Ordinary, incremental

Lean Hybrid Project Delivery

progress can increase team member's engagement in the work and their happiness during the workday.

Remember what we learned earlier in the chapter, make your team members famous!

Method: Sort by Swim Lanes
You may want to move the sticky notes into swim lanes based on the ares of responsibility or specialty to make it easire for line managers to find their current work representations. A swim lane is similar to a flowchart in that it maps out a process, decisions, and loops; however, a swim lane map places events and actions in "lanes" to delineate a person/group responsible, or a specific subprocess. We can make this as coprehensive or as simple as is needed for the successful delivery of the project.

Swim lanes can be vertically oriented or horizontally oriented depending on what you find easiest to work with. The idea of Kanban (a card you can see) is to put the workflow into a visual perspective that makes sense and is easy to use.

Previously in this book we suggested that each leader have a specific color of sticky notes to write on. We can assume that each of this senior leaders is responsible for a specific business unit or group of similar units. These individuals likely have synergy across all her/his domains. At the beginning of the Charette chapter I suggested each discipline have a different color, this then, is why. If the business leaders have a separate color at the beginning, it is easier to create swim lane based on the disciplines they lead.

There are some obvious positives to this method. Among these is, those team memebers who participate in the pull planning sessions can quickly find their work based on the chronological flow from left to right across the wall. Conversly, I have also experienced that it can

Lean Hybrid Project Delivery

be a challenge to tie the prerequistites and subordinate activities or tasks. This is particularly so when a risk has become an issue and the associated activities or tasks have to be realligned to represent more accurately the new timeline. (Thorsten Litfin, 2011)

Tool: Calculating the Critical Path
Critical Path Method (CPM) scheduling is used and accepted on many major projects to plan and coordinate work. CPM works most successfully when the entire organization, from the owner and the general contractor to the subcontractors and suppliers, are involved in the input of information. (Baki, 1998)

Step back and look at the visual representation of the Kanban board after your team has sequenced the activities, after the prerequisites and subordinate activities have been arranged to the chronological sequence and you will quickly recognize the visual similarity to a project network diagram.

This methodology does not tend to need a CPM report however in large strategic or capital projects there are often stakeholders who have a direct interest in where the critical path lies. It is relatively easy to determine this by tracking the prerequisites and subordinate tasks. Since the pull planning methodology work to eliminate as many of the float times in the project as possible this critical path is easy to find.

I prefer to use Microsoft Project as the electronic medium for transfering the data from the sticky notes on the Kanban board. (I will sometimes refer to these as cards.) Then saving the data as an Excel spreadsheet for use in the commitment log, (we will cover this process in the next chaper.) By using Microsoft Project one can use the software to plot the critical path, thereby saving duplication of efforts.

Lean Hybrid Project Delivery

Tool: Future State Validation

Future State Validation builds on the Future State Review of the previous chapter. It's worth while in this phase to revisit that review and validate that the things you thought were the future state remain accurate. If there have been any changes, it is important to note whether ot not they impacted the timeline, the budget, or the quality. If any of these are impacted by the validation process we need to determine if we did something incorrectly (it needs to be corrected) or if we did not "see the future" correctly in the past (which is why we do things this way) and we need to update the leadership (primary stakeholders) whether or not it impacts the budget, schedule, or quality so they remain informed and involved.

Tool: Planning Testing

Testing should be occurring at the earliest reasonable moment throughout the project. We should be planning for testing, developing test scripts, asking team members to assign themselves to the various testing processes (taking ownership) and these should be reflected as sticky note activities on the Kanban board with prerequisites (what needs to be complete before I can test) and subordinate (what tests need to be completed before I can ...) acties and tasks.

Remember: Almost everything has an Information Technology element. And even those that don't probably have some other form of technology applied to them. Unit testing should not be overlooked.

Case Study

Let's use the following example, an existing busy clinic is chosen to open an Urgent Care Center. It is an executive decision and makes perfect sense. However, when the workers begin to make the modifications, they discover the facility is in considerably worse condition than expected and the work will cost much more and take

Lean Hybrid Project Delivery

much longer to complete. At the outset the parties were not collaboratively engaged. The existing managers knew the condition of their clinic, they could have helped inform the construction people and architects, who in turn might have been able to better inform the management as to the scope of work and cost. By keeping all the parties deeply engaged the project manager could have saved the effort of having to "walk it back" several times as the onion being peeled continued to find more and more difficult circumstances.

The parties were not engaged so the project took much longer and cost much more than was originally planned for. Using the Lean Hybrid Project Delivery methodology, we are learning in this book, we might have discovered some of these issues in advance and built-in contingencies and escape plans. Refer back to the Review Risks section in the previous chapter.

In this example the Project Manager did everything correctly as it applies to the common methods of project management. These are examples of what we seek to learn from and improve upon. We know these types of efforts will continue because we are in healthcare and this is how it happens in healthcare, but we can improve and learn by applying the methodologies laid out in this book.

True North

It is never too late to look at the compass and work out what is true north in our projects. As a team we will have to continue to work toward the Lean Hybrid Project Delivery model. This means that starting now, we, as Lean Hybrid Project Delivery Professionals, need to start working together to see where our assigned projects interconnect with the overall target value of the organization.

Does the Mission Statement of our organization give us some level of understanding of true north? Do the Conditions of Satisfaction align with it?

Lean Hybrid Project Delivery

How can we integrate all the projects into this methodology? Now is the time to assure alignment with the Conditions of Satisfaction. Now is the time to find out what is really wanted and the best way to do that is to engage everyone. Who knows if the Conditions of Satisfaction would have changed in the example above if all the parties had discussed everything that was known as well as everything that would have to be done to reach those Conditions of Satisfaction? Maybe the Conditions of Satisfaction would have been negotiated into a different set of Conditions or maybe the requirements of Satisfaction may have changed.

Now is the time for us to think about all the various individuals who are working sometimes unseen by the workflow but whose efforts and the things they build, or support are directly influential to the implementation of or outcome of the project. Let's include them now. Remember it is about workflow and waste reduction every step of the way.

Tool: Pick a facilitator for the next meeting. The facilitator runs the meeting – see a sample of a typical agenda later in this book.

Tool: Set a date for the next meeting if it is not already done recurrently.

Tool: Collaboratively plan the agenda for the next meeting.

Tool: Plus / Delta is the ending ritual of every meeting.

Who are the "last planners" in our project?

7 - Commitments

> **Team Building Part 3**
>
> **Team Building**
> The process of Team Building is constructed on trust. Trust is built by being true to your word and following through with your actions. The point of building trust is for others to believe what you say. Keeping your word shows others what you expect from them, and in turn, they'll be more likely to treat you with respect, developing further trust in the process.
>
> "If everyone is moving forward together, then success takes care of itself." – Henry Ford

Up until now in this book we have covered two subjects broadly. The first is an *Introduction to the Methodology*. This was covered in Chapter 1 – the Introduction to Lean Hybrid Project Delivery and Chapter 2 – the Basic Keys of Success for Lean Hybrid Project Delivery. These lay the foundation for the **mindset** one must develop to be successful with this methodology. The second is *Planning the Project*. This was covered in Chapters 3 through 6. Now, in this and the following two chapters we will be covering *Executing the Project*.

A few years ago, I was exposed to the Lean Construction Institute's "Last Planner" methodology. As a result of that and my ability to look at it from the perspective of the agile/scrum practices that I was already exposed to and using in Information Technology, I was able to begin to formulate the hybrid methodology that is in this book. I would therefore be remiss if I did not credit the Lean Construction Institute's methodology for its influence when developing this hybrid methodology. The term "Last Planner"

Lean Hybrid Project Delivery

is a registered trademark of the Lean Construction Institute (LCI). In this volume we do not adhere strictly to the methodology of the LCI system but are influenced by it to some degree. (LCI, 2017) The goal of the system is to bring stability to the project by giving attention to flow while reducing waste created by the transition of work from one specialty to the next in a construction environment thereby continuously improving the project workflow situation. (Seed, 2017) The Last Planner System of Production Control was developed by Herman Glenn Ballard in his thesis submitted to the Faculty of Engineering of the University of Birmingham for the degree of Doctor of Philosophy, May 2000. (Ballard H. G., 2000)

> "Teamwork is the ability to work together toward a common vision. The ability to direct individual accomplishments toward organizational objectives. It is the fuel that allows common people to attain uncommon results."
>
> Andrew Carnegie

In his thesis Doctor Ballard states, "This thesis extends (The Last Planner) system application to those coordinating specialists, both in design and construction, ... one of which also explores the limits on unilateral implementation by specialists."

The LCI System produces and expands predictability of project workflows. Results are the outcome of people working together in iterations of planning discussions that produce a "network of commitments" based on the Kanban like planning methodology they use. This methodology is essential to identify the work, tasks, and products needed to accomplish the project milestones. This network of commitments makes work output (product) ready and assures a **person** has promised to complete it. All the while, we are learning

Lean Hybrid Project Delivery

from our experiences gained in the iterative processes. This must be accomplished in a collaborative environment. Commitment and responsibility are two-way agreements! In the Lean Hybrid Project Delivery methodology, we build on this proven technique to deliver our projects on schedule and then budget.

Building on what was completed previously in the Flesh Out and Sequencing chapters, we continue in this set of iterative meetings to create an environment of personal commitments. This is not necessarily a published schedule or project plan, it is the supporting instrument or tool, for engaging the project team in completing work consistent with the deliverables promised for the project. (Greggory Howell, 2005)

Method: Building Commitments
The Lean Hybrid Project Delivery method can be tailored to project circumstances, and we find it very helpful in Healthcare, Information Technology, Construction, and Systems Integrations projects because the intention of the system and the fundamental nature of the practices involved are clear: "Produce predictable uninterrupted workflow by creating a coherent set of commitments that connects the work of the specialists to the promise of the project to the client and coordinates their actions."

This happens in five iterative "conversations" designed by Dr. Ballard so the team can manage the network of commitments inside each of their accountabilities and specialties. During the Charette, Flesh Out, and Sequencing chapters, we built a Kanban board of interrelated activities and tasks, we prioritized them, we sequenced them, by reviewing and analyzing their relationships, to each other and to the overall project. We also assigned various individuals the responsibilities of accomplishing those tasks or activities and finally we assured that the chronology of the overall project was accurate by assuring that we had identified the hard constraints relative to the various activities and tasks. Now if we step back and look at the

Lean Hybrid Project Delivery

Kanban board, we have a visual representation of the activities as they flow through the project. We have a visual representation of a network diagram of the project. We have a network of commitments. However, if the individuals who are assigned to the various activities or tasks have not taken ownership of those activities or tasks because they were assigned to them not taken ownership by them then we have not gotten commitments.

This is the purpose of this group of iterative meetings. In this group

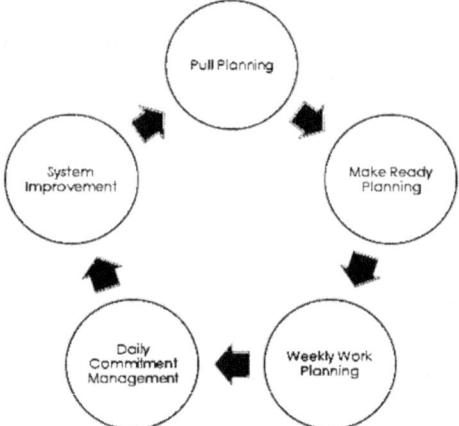

of meetings, the individuals to whom the activities or tasks are assigned are participating collaboratively among themselves and along with us, and their leaders, to assure that the estimates given in the Charette, the Flesh Out, and the Sequencing chapters were accurate.

Refer to the graphic above. This is my translation of Dr. Ballard's workflow. It illustrates the iterative activities associated with the set of meetings defined by this chapter.

Method: Pull Planning.
As Lean Hybrid Project Delivery Professionals, we facilitate the team of people who are ultimately responsible for delivering a project milestone. We help them plan together how that will happen. The **planning conversation** produces an integrated project delivery plan

Lean Hybrid Project Delivery

for what will be delivered at each hand-off. The hand-offs are defined by the team members during the Charette, Flesh Out, and Sequencing meetings, but also during the Commitments meetings that can also be referred to as sessions or conversations. Pull Planning meetings (or sessions or conversations) are exercises whose output is a schedule for those hand-offs. This schedule prepares the team for action together. The future is not set and remains uncertain.

Beginning with the Charette, continuing through the Flesh Out, and culminating in the Sequencing; the plan developed is based on a series of requests starting from the final milestone and working backwards – **the pull**. This group of iterative meetings is finally bringing that group of commitments to the GEMBA (the workplace). Here, in this the Commitment chapter, is where we are planning the SCRUM or iteration by week (or two weeks) to accomplish the tasks. Here, the individuals are taking ownership for the actions. Each of these requests form the basis of the individually identified and personally promised task or deliverable that translates into the network of commitments necessary to deliver the phase and establishes when the work will (is promised to) happen.

Therefore, there are four pull planning conversations, if the project is large enough. One of these would be to pull the various milestones of a larger project in the Charette. Then the pull planning sessions can be performed to individual milestones in the Flesh Out and Sequencing. This chapter only deals with pulling to a single milestone in a single time period. Here we are not focused on the entire project. We are only focused on the next period of time, that can be six weeks, a month, two weeks, or a week, with a preference to the shorter time period.

The Lean Hybrid Project Delivery Professional facilitates this conversation but the actual tasks, durations, sequence of events, and assignments are accomplished by the team members themselves. The team is making promises to each other, not to the Project Manager. This is completed in this series of conversations, but often, due to the time constraints imposed on the business is actually performed in the Pull Planning conversation session.

It is important to remember that many of the individuals who are taking responsibility for these tasks and activities have other primary responsibilities or jobs in the company. The project is usually something extra for them to do. So, in the first few iterations of this set of meetings or sessions it is important to work with them to help identify accuracy in their estimation of what they can complete per iteration or sprint. They will get better as they practice and as they learn to develop as a team.

Method: Make-Ready Planning.
Each team member has graduated from interested observer to responsible individual for a group of actors (crews, teams, and individual persona) and reviews the work in the coming six weeks look-ahead period. She or he identifies whatever is needed to accomplish the work and makes the requests and receives promises (personal commitments) needed to assure the input or requirement will be available when essential.

Most (if not all) of these tasks will have been identified in the pull plan, minus the durations and sequences. In this conversation personas ask for help whenever they lose confidence that the work will be ready when required. They know their part of the work but there are other workstreams that must also be working to create a Lean Workflow.

The team adjusts or reaffirms the plan in regular meetings during the lookahead period. No single individual is responsible to the whole plan, but every single individual is responsible to the team to accomplish their part of the plan. The make-ready plan establishes what can happen and serves as the basis for securing reliable commitments for the coming scheduled work. The commitments are made from the line manager and her or his team in each specialty to other line managers and teams in other specialties. These are they referred to as the last planners in Dr. Ballard's method.

Tool: Planning for external constraints
The Make Ready planning method is designed to look down the road far enough that the planners may anticipate lead-times outside of their control. An example of this would be orders and durations of times for those orders to arrive. For instance, if one needs to place a computer order and those computers must come from China, then placing the order needs to occur with sufficient time for the shipping from the Orient to complete. Additionally, one must be advised of the durations of times that things outside of their control such as procurement processes and purchase orders may entail. Internal corporate processes must always be accepted and anticipated. I would recommend that the procurement staff be included in the team.

This is also where you would be thinking of the detail of compliance with regulatory bodies and other governmental concerns. These constraints must be calculated in your project and addressed in this step.

Method: Weekly Work Planning.
Final coordination of the work in the coming week (or two) is completed as each last planner makes promises to the other last planners and the project manager asserting what will be delivered, to whom, and by when. Capacity, resources, and finances are allocated by day in support of those promises. These conversations promising deliverables occur in a group setting. This meeting allows other last planners and team members to assess consistency with their own promising. As a result, adjustments to the weekly work plan (WWP) are made during the meeting.

Lean Hybrid Project Delivery

Method: Daily Commitment Management.
Like its cousin the SCRUM meeting, (Layton, 2012) this is a formal (not a hall meeting) however a brief "stand-up" meeting (or conference call). The appropriate members of the team are involved and are held daily to report WWP completions and complications. During these meetings the team lead, line manager, or last planner works with the team to adjust to the circumstances, re-promises as necessary to the project team, (through the Project Manager). Personas also get help from each other and record plan variances and their reasons. This is the root work meeting and serves a self-directed controlling or steering purpose that allows those closest to the work to adjust to the always changing conditions of the project, so they can complete all their promises. (Greggory Howell, 2005)

> Practice the KISS principle. Most people like to define this as Keep It Simple Stupid ... I prefer to turn it around a little; Keep It Stupid Simple
>
> Timothy Tuohy

The Lean Hybrid Project Delivery method drives planning reliability to continually improving levels.

Tool: Line Manager Kanban
The line managers may find it helpful to have their own Kanban board in their respective work areas. However, this Kanban should be a simple Lean Kanban. This is a board (or a panel on the wall - or whiteboard) is typically divided into four columns. The first column is the "backlog". The backlog is the list of all the tasks and/or activities to be accomplished in the next sprint or in this case the next week. The second column should be "specify work" or "work in progress" - depending on the scope of the work involved. For

Lean Hybrid Project Delivery

example, if there is a column that refers to specify work this indicates that there is additional planning to be done such as the layout on the floor where the studs would go for a wall. Work in progress then, using this example, would be building the wall. It may not be necessary in the local Kanban to have a "specify work" column at all. Practice the KISS principle, don't do any more overhead work than is necessary. It may be all that is needed is a work in progress column.

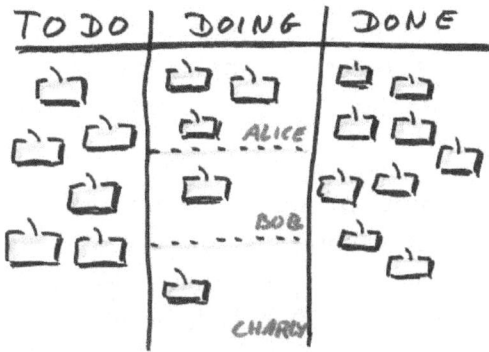

The next column would be "implement work" or "install work" if one is fabricating something (such as HVAC Ducts) or in the case of the construction example above there would not be an implement work - there would only be a "completed work". In some cases, these simple Kanban boards also include a "validate work" - which implies that someone goes by and validates (or checks) that the work was completed properly before the work is considered to be finished. The point being, this Kanban is simple. It is there to show work that has not been completed, work that is in progress, (and who is doing it) work that is complete, and if necessary, work that has been checked and who is checking it.

Tool: Pull Planning as the Root

Let's take some time to deeply explore pull planning because it is so very important to our success. Pull planning is an intentional retrospective review of the final steps of the processes that lead to a conclusion. Pull planning is a decomposition process. We are

Lean Hybrid Project Delivery

imagining ourselves going through time backward from the point of the finished product or output.

All of us has performed a postmortem on a project that has gone poorly hoping to determine how we might have done better or, in very bad cases, how things went totally off track or even failed. We have had the dubious honor of taking the blame for poorly planning something that did not work, and the resources had to scramble to remediate the process, so the project would work. We have performed postmortems on project implementations that had to be backed out because they didn't work or worse yet, broke something else that was working.

One method we may have attempted to eliminate those results (particularly in very sensitive projects) is a premortem. This is a practice that starts by saying the project has failed, now come up with all the reasons you can think of that could have caused it to fail. (Portny, 2013) Not all retrospectives are from a negative perspective. However, all retrospectives will help us avoid negative outcomes.

We're spending a lot of time on this because we sometimes have to help the last planners understand what we are helping them do. In the third chapter we created the Conditions of Satisfaction. Those Conditions of Satisfaction describe what the last planners are ultimately working to deliver as we apply workflow analysis. We are looking for ways to reduce "T" as it applies to "L" in every activity in the project delivery.

$$L = \lambda * T$$

(Work-in-Process)   (Throughput) *   (Lead Time)

Lean Hybrid Project Delivery

We may want to begin by reminding them of the last go-live they participated in and ask questions like:

- What was the most challenging thing you overcame when the product went live? How could we (as a team) have done better?

- What was the last contribution you and your coworkers made to the product on the day before the go-live?

- What did you and your coworkers have to wait for from another team or vendor before you were able to perform the last task before the go-live?

We may also employ the five-why method. We are decomposing their past go-live. Once they begin to understand, we can employ the transference technique (Freud, 1895) but instead transfer the thought process from a project that has occurred in the past to a project that has not yet occurred. This will become easier the more frequently we and the other team members practice it.

We as Lean Hybrid Project Delivery Professionals are looking for the hand-offs between groups or teams. No team or individual does all the work in a project by themselves. No work is done in a vacuum. All project work performed in a vacuum fails. (Harvard Business Review, 2012) In our last chapter we identified specialty swim lanes, in this pull planning conversation we, as Lean Hybrid Project Delivery Professionals, are looking for the "links" where a product migrates from one specialty swim lane to another, we refer to this process as a hand-off. Hand-offs are critical to the success of the pull planning methodology. We begin by looking at the go-live and then looking backwards in time to what the last hand-off was prior to the go-live. The term go-live actually refers to the final hand-off where we, as information technology professionals, handoff the product to the end users. (LCI, 2017)

In the pull planning conversation, our product is a network of

Lean Hybrid Project Delivery

commitments. Personal commitments are made between the party that is handing off and the party that is receiving the handoff.

Humans are notoriously poor planners from the rear. We have all heard the saying, "hindsight is 20-20". So why not start at the beginning looking backwards? By assuming the product is complete and decomposing the components and tasks that were required to complete the product, and by doing this collaboratively with all those parties that are associated with developing the components and performing the tasks we can better plan as a team.

It is important for us as Lean Hybrid Project Delivery Professionals to fully internalize the ideology of Lean Hybrid Project Delivery. It is a very agile concept to include all the technical players of product development and composing those technical players into small teams that develop in sprints. One of our tolls is an application of the Last Planner System (Ballard H. G., 2000) is a cousin to the agile methodology but expands the idea of teams to include the end users and stakeholders in the planning. We do this because we are no longer project managers but facilitators of the team of individuals completing a project as a team. In larger projects more than one team or many teams may be working simultaneously to accomplish the overall project. The big room meetings that we have every two weeks or month calls in the leads, line managers, or last planners of those teams. By forming this team of teams, (McChrystal, 2015) we can then work collaboratively to fine tune the team hand-offs so that the plan that was originated in the first meeting or conversation can continue to become more accurate and more cost-effective. This is the very heart of Lean Hybrid Project Delivery. (Womack J., 2003) (Ballard I. D., 2016)

In Lean Hybrid Project Delivery our goal and our mission is to facilitate a team or group of teams in their efforts to fine tune the design, methods, and plans of their project so that the least amount of work may be accomplished to efficiently and effectively deliver the

Lean Hybrid Project Delivery

project on time and within budget. Among the most important things to focus on while "Leaning In' is the lead times and the hand-offs. Due to the iterative conversations among team members that we are facilitating, constant incremental performance and planning improvements are occurring within the team or teams. (McChrystal, 2015)

Employing the tools above and bringing as many of the line managers or last planners into his big conference room as we can find, push all the tables out of the way, move the chairs out of the way, and get everyone up on their feet.

Pro Tip: Everyone participates in pull planning.

As the facilitator of this hybrid project delivery, you should bring some things with you to this conversation. You should bring colored sticky notes, a different color for each specialty swim lane, and felt tip pens, enough for everyone. Distribute the colored sticky notes to each person who is contributing for their specialty. Each of the sticky notes is organized as shown in picture below. In the upper left corner of the sticky note are the initials of the individual who is responsible

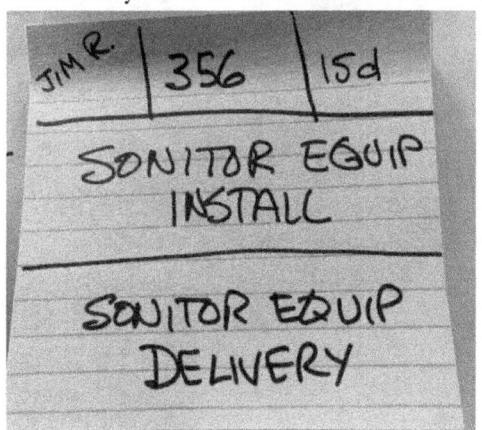

for the individual task that is listed in the center of the sticky note. At the bottom of the sticky note that individual should write a description of anything they need someone else to accomplish before they can perform the task that they have written in the center of the of the sticky note. In the upper right corner of the sticky note the duration in days of that task is written.

Lean Hybrid Project Delivery

The sticky note represents an individual's commitment to fulfill an individual task in a duration of time. The section at the bottom of the sticky note which stipulates a requirement the individual has in order to accomplish the task that they have committed to should have a matching individual on a different sticky note who is committing to deliver that deliverable. As these individuals write the sticky notes, they should stick them on the board in the order they initially think they are accomplished.

Remember, we are working backwards from the milestone, so the first sticky note on the board should be the last task performed before the go- live. The next sticky note on the board should be the task that enables or informs the last task prior to the go-live. Using this example, we should have

> Problems are a sign of life
> Norman Vincent Peale

a diamond on the far-right side of the board labeled "go live" and to its immediate left there should be a color sticky note which has an individual identifying the task that they are going to complete with the duration of that task, and that should be the last task before the go-live that enables or informs the go-live. And the next one to that enables or informs that task.

This exercise is repeated over and over as each of the individuals who are standing in the room as part of the specialties that are required to accomplish the product or service begin to decompose the tasks from the go-live to the inception.

Process: Building Trust
As this process continues, we begin to realize that we are building a network of commitments. Making and keeping commitment is the basis of trust. (Heather Craig, 2021) It's not important at this part of the planning that the sequence of events is perfect.

Lean Hybrid Project Delivery

What is important is that all the activities, or tasks, are identified. In this iteration of the discussion all the members of the team, even though they are from various specialties within and outside of the specific field we are supporting, are talking and working together to identify all the tasks that must be completed to reach the milestone. As a Lean Hybrid Project Delivery Manager our role is coaching and encouraging these individuals to dig deeper into the planning concepts and identify the tasks that they not only need to accomplish but need to have accomplished so that they can accomplish their tasks.

> Building Trust – Tip 1
> Avoid self-promotion
> Heather Craig

When we feel this iteration of the conversation is beginning to wind down, it's time to interrupt the thought process of the individuals and start reviewing the sequence of the events. Again, we are the facilitators not the sequencers. The team now begins to sequence their own tasks. We are only facilitating them in making their own plan. No longer are we the task masters or task managers, but we are facilitating those individuals who are performing the tasks within the team so that they may identify the work that they must do and the work that the team must do in order to accomplish the task. Once the sequence has been set by the team, we validate the individual commitments that have been made on the sticky notes. We will have noticed during the process;

> Building Trust – Tip 2
> Be true to your word – Honor your commitments – Don't make promises you can't keep
> Heather Craig

Lean Hybrid Project Delivery

individuals will have adjusted the durations that they believe it will take to get their jobs done because they are now working directly with other individuals for which they have been compensating unknowingly. Durations for individual tasks committed to by individuals during the pull planning process tend to get shorter with every iteration of pull planning. As the individuals become more comfortable with the technique and tools of the pull planning the process is building trust among the team.

> Building Trust – Tip 3
> Communicate Effectively
> Heather Craig

As Lean Hybrid Project Delivery Professionals, we are no longer guessing what the durations were going to be or where the hand-offs are going to happen because we have determined both the hand-offs and the durations of those tasked with the entire team. We have facilitated the team in their ability to do their job effectively efficiently and in a timely manner. We have found through experience that early resistance quickly falls away as the team members begin to understand the value of the pull planning conversation.

Remember 90% of our job is fostering effective communication.

> Building Trust – Tip 4
> Build Trust Gradually
> Heather Craig

Once this is complete, take the time to ask the participants if they have found any value in this conversation. One method for doing this is called a "plus / delta". This is a simple process in which standing at the board, we write a big plus on the left and over to the right we draw a triangle. Asking everyone in the room to participate, we ask

Lean Hybrid Project Delivery

them to identify things that we have done well in this meeting. We write that list under the plus. Then we ask everyone to identify things that we as a team could have done better, we write those items under the triangle. In this manner, the entire team participates in the lean practice of constant improvement as well as learning from each individual event as opposed to waiting until the end when we don't remember everything that happened during the project and it's therefore difficult to learn from it.

> Building Trust – Tip 5
> Make decisions collaboratively
> Heather Craig

This concludes the pull planning conversation or meeting. However, it does not conclude our work as Lean Hybrid Project Delivery Professionals. It is our opportunity now that the team has identified all their tasks, committed to each other personally, estimated the durations of their tasks, and sequence the events - to create a commitment log. We found the easiest way to create a commitment log is to do so on a spreadsheet. There is no set rule of how the spreadsheet should be created however, the fields that are on the sticky notes as a minimum should be converted into columns on the spreadsheet. Using the current date and the durations provided by the individuals as committed to each other, it is relatively easy to produce a series of dates based on durations relative to the current date. By doing this, the Integrated Project Delivery Professional can quickly produce a

> Building Trust – Tip 6
> Be consistent
> Heather Craig

Lean Hybrid Project Delivery

timeline for the anticipated completion of the project.

Weekly, we should hold a 15-minute standup or teleconference meeting to review the commitment log. The purpose of this meeting is very agile in nature. We are not holding the meeting to beat people up, but rather to find out how the team can support each other in accomplishing their goals. As the facilitator we may find that an individual has stopped forward motion on their tasks because they have run into a constraint. It is our job as project facilitators to knock those constraints down so that the individuals may accomplish the tasks that they've committed to.

> Building Trust – Tip 7
> Participate openly – in team settings, show your willingness to trust – Listen actively – give feedback respectfully
> Heather Craig

Pull planning is an iterative process. It is not a "once and done" exercise. If we can leave the sticky notes up on the board where we're working – we should do so, but if we can't it's worthwhile to find a manner of moving those notes in their correct order onto some type of media that allows us to move them from one point to another in case we need to have our next pull planning conversation meeting in another room. This can be done on roll paper that is readily available in places like Home Depot, or it can be done on poster board or can be done with a photograph from our cell phone and then move the sticky notes onto some paper that allows

> Building Trust – Tip 8
> Be honest – Always tell the truth
> Heather Craig

Lean Hybrid Project Delivery

us to keep them in order. One manner by which the project manager can help themselves maintain the order by which they move the sticky notes from the board is to use the center part of the upper section of the sticky note to number the sticky notes for location sequence.

> Building Trust – Tip 9
> Help people – be kind - Show your feelings
> Heather Craig

The commitment log that we have created should either be posted in a commonly available place such as a shared network drive that all members of the project team have access to or e-mailed to all members of the project team.

As soon as we know there is a project, we should start assembling the team of last planners and stakeholders. We should include everyone who contributes to, has a deliverable in, finds value from, and is affected by our project when we are seeking with whom we should collaborate in our project.

When we have the definition of the final product we should begin planning retrospectively from it. That definition is our **true north**.

> Building Trust – Tip 10
> Admit mistakes
> Heather Craig

Right now, we should start reprogramming our minds from the conventional waterfall project planning forward methodology to that of Lean Hybrid Project Delivery! No single project methodology is sufficient to successfully deliver a large complex capital or strategic process. We must be willing to use any applicable method within the construct of the project delivery model.

Method: Weekly Work Follow-Through
If there is a nitty-gritty to go with the nuts and bolts, the weekly work planning section is the nitty-gritty. This is where we find out how reliable our promises and commitments have been to each other. This does not imply that commitments made were made falsely as much as it says that we are not necessarily good at estimating when we start out. Each iteration of pull planning and make work ready planning fine-tunes the commitments and timelines and therefore the expectations of the project. Healthcare Information Technology projects traditionally encompass multiple disciplines and specialties, each typically comes to the project with independent and mutually exclusive goals. These differences can lead to misunderstandings that in turn lead to incorrect work being done and therefore multiple reworks within the project. Each of us as Lean Hybrid Project Delivery Professionals has been confronted with the need to back out a go-live because it either didn't work or it broke something else that was already working.

Lean Hybrid Project Delivery empowers performers to offer conditional promises or commitments which leads to negotiation. We are very familiar with these conditional promises in Information Technology. In coding we have a conditional branch we refer to as "if, then, else". This, in its own way defines conditional promising. In his book "Conversations For Action", Fernando Flores modeled basic action workflow as a series of mutual promises between customers and performers. (Flores, 2012) In his model of the **basic action workflow loop**, each individual workflow is a structure of commitments that constitutes transactional agreements between two parties within the loop. What we have created through the previous chapters of planning is a practical "network of commitments". If we stand back from the wall where we have performed the pull plan

Lean Hybrid Project Delivery

what we see is a "project network diagram"! (Project Management Institute, 2017)

Lean Hybrid Project Delivery projects tend to be a complex matrix of systems that work together, interact, and have multiple integrated budgets. Many decisions are made when developing and delivering projects to meet the Conditions of Satisfaction. No one professional, nor even a small group, can make all the necessary decisions to produce a successful project outcome. We are confronted with too many issues, technical and medical skills, inputs, software knowledge, options, and regulations. A Lean Hybrid Project Delivery project uses Cluster Groups to better manage this task.

Tool: Cluster Groups

Cluster Groups are small groups of individuals assigned to a related task, usually grouped by specialty. These are similar to the agile **scrum team**. The daily and weekly work units are performed in these Cluster Groups. (Since we are Lean Hybrid Project Delivery Professionals, feel free to use the term scrum team if you're uncomfortable with the term Cluster Group.)

Cluster Groups should be formed around whatever grouping is appropriate to the project. The cluster group should collaboratively form to define objectives and innovations that it will bring to the project.

As with the agile scrum team, daily huddles may prove to be of great value as a tool for the cluster team to plan their daily work and assure the project schedules, individual commitments, and output products are completed on time. Daily huddles are not necessarily problem-solving or planning meetings. Rather, they are the means for the team to surface unresolved issues collaboratively. These should be standing meetings, as in *standing up*, to force the meetings to stay short and agile. (Agile Alliance, 2017)

The objective of this phase is to focus on effective, efficient, quality

Lean Hybrid Project Delivery

work. This daily huddle helps to maintain workflow progress, identify constraints, measure performance, and learn from variances in the plan. (Seed, 2017)

Tool: Weekly Check in Calls or Meetings
Weekly check-in calls, of the duration no longer than fifteen minutes, along with daily huddles at the Gemba or **Locus Opus,** are the root of this cycle of Lean Hybrid Project Delivery. If you've been exposed Information Technology, you may be aware of the OSI seven-layer model. The Lean Hybrid Project delivery model employs the Last Planner layered model which is similarly constructed. At this layer the work effort is referred to as **"will". We will do ...**

The question that is answered at this layer, in this conversation, is:

"What will we do?"

There should be a limited scope of work effort being planned in this phase. The individuals who are now preparing for their task delivery should be looking no more than two weeks in advance.

The two questions that are being asked and answered in this phase are:

1. "What do I need to do to accomplish my deliverable task in the next 1 to 2 weeks?"
2. "Is there any obvious constraint that needs to be overcome or eliminated in order to accomplish my deliverable task in the next 1 to 2 weeks?"

The answers to these two questions sum up the list of things to do for the individual performers over the short duration look ahead period.

This phase or layer follows the Make Work Ready phase or layer, in which, the questions was answered – "What **can** we do?" Therefore, in the previous layer or phase the team has already broken the

Lean Hybrid Project Delivery

workflow down into digestible bites, or chucks so that the Cluster Groups can efficiently complete the work. These chunks should be no longer than two weeks duration.

Tool: Reporting Percent Plan Complete
Learn and practice the performance of Percent Plan Complete this helps gauge the reliability of the planning system as we apply it. Percent Project Complete is the number of planned activities that we have completed divided by the total number of planned activities. The outcome is expressed as a percentage.

Percent Project Complete measures the extent to which the last planners (line managers and supervisors) commitment (in the WILL segment) was realized (Ballard H. G., 2000).

Project Percent Complete tells us if the planning process is reliably predicting what will actually be completed. This tool's output can be misleading if the tasks completed exceed a duration of two weeks.

This informs the Monitor and Control element of the Lean Hybrid Project Delivery method. Please see the next chapter under the section "A3" to see how this fits into reporting the overall project status to management.

Tool: Weekly, daily Activities.
Weekly: As with a scrum team the Cluster Group is working together to affect a change or build a product. A typical week would be similarly modeled.

Monday – Morning check-in session (15 minutes) formally called, stand-up, and on the calendar.

Wednesday – Morning check-in session (15 minutes) formally called, stand-up, and on the calendar.

Wednesday – Afternoon (as needed, minimum every two weeks) pull-planning session against the plans from the Big Room. Review of the

Lean Hybrid Project Delivery

Constraint Log, review variances.

Tool: Perform Project Percent Complete.
Friday - Morning check-in session (15 minutes) formally called, stand-up, and on the calendar.

Daily applying the Deming PDSA methodology of systematic learning and continually improving our products, processes, and service. (Deming W. E., 2017) This is accomplished in the form of daily huddles. These huddles should take place as close to the location where the work is actually happening as possible. Lean

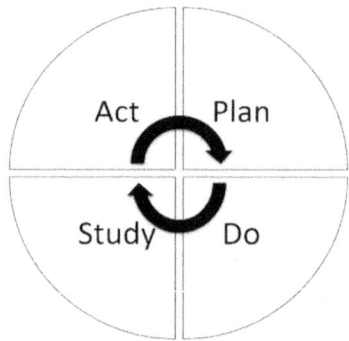

practitioners often refer to the work location as the "Gemba", and again the Latin is "Locus Opus".

A retrospective learning event should occur with the completion of every deliverable or monthly to learn how we could "do this better".

Lean Hybrid Project Delivery

8 – Monitor and Control

Monitoring and Controlling

Managing to the Desired Outcome

The processes of monitoring and controlling projects do not belong in a chronological location of the process of project delivery. They are interwoven into every part of the project delivery life cycle. This chapter delineates these processes, methods, and tools for you to be able identify them as they are occurring.

"If you focus on principles, you empower everyone who understands those principles to act without constant monitoring, correcting, or controlling" – Stephen Covey

There are advantages in this practice. One of the primary advantages of this practice is allowing the executives and end- users, those people who will own the product once the project is complete, to understand both the work effort and the cost of accomplishing their dream.

An ultimate milestone described by the person that desires it without complete comprehension of the cost or work effort required to accomplish that outcome is best described as a dream.

In all practical terms the <u>dream is a good thing</u>. There is no rule that specifically states that a dream is vaporware. Without a dream, without a vision, there is no advancement. We as Lean Hybrid Project Delivery Professionals thrive upon producing the miracles that end up making dreams happen.

The iterative process of monitoring and controlling is not designed to talk the dreamer out of their dream. Rather, it is designed so that all the participating management parties may

Lean Hybrid Project Delivery

have input into the milestones that will have to be accomplished in order for that dream to be realized. The realization of that dream may require innovation, imagination, and new design.

Through the iterations of this process, it may be determined that the dream is entirely reachable sooner, with less effort, and less expense. This process may also reveal that the expected work and expected cost are exactly what the dreamer expects. Of course, it could go the exact opposite way to be determined that this is much costlier, take much longer to complete, and may not even be a realistic dream. We may not have the technology to accomplish it which will require additional innovations.

This is the process where we are *executing* the document that is the description of the specific expectations of the stakeholders as to what are their Conditions of Satisfaction.

Monitoring and controlling the project is integral to the execution of the project not external to the project. It is not about watching what is happening and reacting to it, nor is it about directing the outcome. It is about coaching the team and teaching the principles in the preceding chapters.

> "If you focus on principles, you empower everyone who understands those principles to act without constant monitoring, correcting, or controlling"
> — Stephen Covey

Process: Ask Questions
Let's tie it all together, you will see there is nothing new in this chapter. I am merely tying it all together is a neat, actionable package.

Lean Hybrid Project Delivery

In every iteration from the first minute of the first iterative meeting to define the Conditions of Satisfaction through the close out of the project, questions are the root of communication and integral to the success of the project delivery. While sitting around the table in the big room we, as the Lean Hybrid Project Delivery Professionals, realize our new role. Now our role is that of facilitator. Facilitators, in this process ask questions.

There are many questions that we as Lean Hybrid Project Delivery Professionals have learned to ask through our own experiences. The difference is we DO NOT DESIRE to be the expert in the conversations or meetings or iterations. We are approaching this to help our clients answer a simple set of questions:

- What will you give?
- What do you need to be given?
- When will you give it?

Even if we know the answers, we pretend to be ignorant. Because we want the individuals in the team to personally commit to each other. We are facilitating **their work** effort, not doing the work for them.

You will notice that the question of when is included in this list. We will come to that question as we **pull** this plan from the milestone backward in time to its inception. Our focus is to eliminate waste (in the form of excessive lead times) as we go. Our goal is to reduce the duration between handoffs by helping determine where we can make handoffs more efficient.

We, as Lean Hybrid Project Delivery Professionals are looking for a complete description of what the actual activities to reach the successful outcome include ... what that <u>final milestone</u> looks like in terms of actual work effort to achieve it.

In this every iteration, we are looking at cost. We are looking at the

Lean Hybrid Project Delivery

scope, keeping the project on true north. We are looking to tighten up the schedule. We are questioning quality, to assure it is included at every step along the way.

We are building a final milestone. At the beginning that milestone is still a roughhewn stone, it does not reflect the end result that will describe the final milestone. With each iteration, we are better

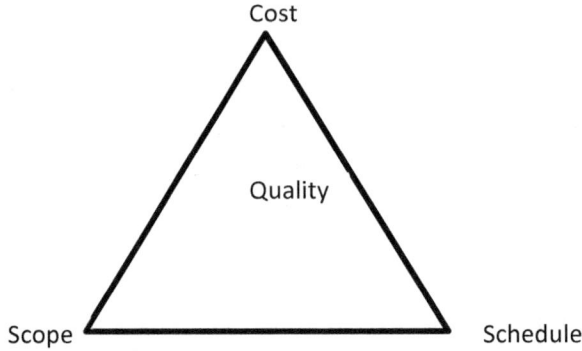

defining the milestone. We are better developing the cost and schedule as we assure we remain focused on the outcome that was defined by the Conditions of Satisfaction. We are both monitoring and controlling the project using the tool "Asking Questions".

Tool: Pull Planning
Now comes the fun part! The pull planning tool is an interactive tool. Get everybody up out of their chairs and have them stand at the board with you. Provide each of the associated parties with a 3 x 5 for a 4 x 4 sticky note pad on which they can begin to write the milestones and activities that add up to the completion of the final milestone. (We need to bring some felt tip markers to the meeting as well.)

When we are looking for the milestones, it is very important that we clearly explain that we're not looking for the tasks that are required to accomplish the milestone but rather the <u>milestones</u> for each of their individual disciplines and specialties that must be accomplished in

Lean Hybrid Project Delivery

order for the final milestone to 'go live'. However, be aware the team will discover milestones as they are brainstorming tasks in the Charette as well as when they are adding activities and chunking them down in the Flesh-Out and Sequencing. Remember we are working to chunk the activities down to work effort that can be completed by an individual or a small team in 5 days or less. The previous chapters have detailed this, I am including this here to clarify that Pull Planning is an integral part Monitoring and Controlling the project.

Applying this tool iteratively through the conversations and meetings we are not only looking for the duration of time that is required to get the job done, but we are also working to reduce schedule through reduction of waste, in this case, time and inefficiencies in hand-offs. That's done by applying a set of principles.

It is the **pull plan** that pulls the tasks backwards from the milestones and also adds more accurate durations to the timeline. Estimating improves with every iteration as does reporting of percent complete.

If your project requires a specific end date it makes this work effort more challenging because you are working within a fixed maximum duration as well as with fixed resources. Hopefully these iterative exercises have given enough high-level decision-support material for the future owner to understand the scope of the work and therefore be willing to negotiate a more realistic desired due date if it is necessary but don't count on that.

Tool: Making Commitments
As Lean Hybrid Project Delivery Professionals, we are facilitating teams of people, individuals, in accomplishing their own projects. We, as Lean Hybrid Project Delivery Professionals are helping them fully understand the work that they and their teams will need to perform.

One of the key processes in these iterative **pull planning** session

Lean Hybrid Project Delivery

meetings is to begin forming personal relationships between the individual performers. We are doing that by getting these individuals to <u>individually commit</u> to each step of the process that must be accomplished, not to us but to the individual who is responsible in the next hand-off.

This Charette meeting is where we begin that process. Each of those people, the <u>individuals</u> that are in the meeting, are going to be asked:

- "What is the activity you must complete in order for your specialty to hand-off the work to the next specialty or the end user?"
- "Can you and your team accomplish this activity?" That's followed by a similar question,
- "Will you and your team accomplish this activity for the individual you are handing off to?"

In the next set of meetings, we sometimes call these meetings "conversations", we begin working with the last planners, that is,

those members of the above teams that are the individuals responsible for planning the actual work that is required to accomplish the milestone. Usually this is the line manager. Be sure to ask the leaders in this conversation for those people.

Tool: Make Ready Planning
After the Pull Plan conversation the last planners, or line managers, return to their respective specialties and plan to get the work done. In construction that often means getting all the materials ordered and delivered so the work can actually begin.

Lean Hybrid Project Delivery

Similarly, in Information Technology there is a need for those who have just committed to getting a task (or set of tasks) complete to produce a detailed design (plan) for how they are going to produce, build, or implement their commitment. In the same method as Agile Project Management, we are helping these technical specialists "chunk down" the tasks into smaller deliverable "sprints". Although the Agile term "Sprint" refers to a team coding effort of delivering working software, a similar workflow methodology applies here. Although the team may be doing anything from determining the network infrastructure changes and additions needed to actually developing code to meet the Conditions of Satisfaction, most often the work effort is implementing an upgrade, enhancement, or new module of a vendor supplied system.

In this phase the detailed specifications of the desired project deliverable are gathered and the individual performers who will be performing the work are determining if they have everything they need to accomplish the deliverable task output as well as when they will be able to schedule it. User group meetings can also be helpful in this phase to determine the specific needs of the delivered product. (Ballard I. D., 2016)

An important element of Make Work Ready phase in the Lean Hybrid Project Delivery methodology is discovering what needs to be ordered, when it needs to be ordered so that lead times do not impact deliverables, and how to pay for orders so the project completion is not affected.

Make Ready Planning is a daily, weekly, or biweekly formal event completed by the last planner (or line manager) and the team associated with that individual. Everything about Lean Hybrid Project Delivery is associated with team building, team commitment, team performance, and team delivery. Each individual is working toward their best performance based on their commitment to the

Lean Hybrid Project Delivery

team and the commitments from the other members of the team that they too are working toward their best performance. Always working to learn and be better.

Semper Melius! Always Better!

Tool: Kanban

In one environment we may a have a Kanban board. While Kanban means a card you can see it is a flexible tool that can be used in many variations. In pure lean it could simply be a camera looking an inventory so that one can know when to order more stock. In this

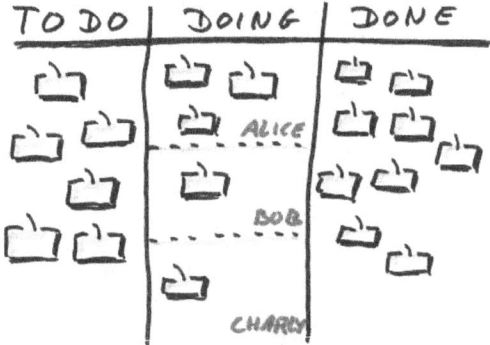

case, I am going to suggest a simple white board in the line manager's office. The use of sticky notes is encouraged as this is easy. Divide the white board into three sections: To-Do; Doing; Done. In Agile and Scrum the term To-Do is referred to as the "Backlog". Here, the line manager's and her team are 'brainstorming" their own list of activities required to accomplish the tasks or activities the Line manager has committed to. This is a local function and is not attended by the Lean Hybrid Project Delivery professional unless he or she is invited to attend and then the attendance should be as a coach not as a manager.

There is a second meaning of Pull Planning in Lean. This is actually the first definition of Pull Planning in the Lean environment. Here, the team is identifying the To-Do tasks and the individuals committing to do them "pull" the tasks from the To-Do column

Lean Hybrid Project Delivery

right their name on it the sticky note and move it to the Doing column. In this manner everyone on the team knows what needs to be done and what is being done and by whom. When complete the sticky note is moved to the Done column.

As Lean Hybrid Project Delivery Professionals, we are employing management by walking about in this phase. We are not actually doing the work. We do need to assure that any constraints that are encountered by the team are knocked down. Particularly at the beginning of this phase as the individuals who are part of the pull planning session are learning how to implement the commitments they made in the pull planning session. Our roll now should be facilitating their understanding and learning. (Brechner, 2015)

There is a fine line between facilitating learning and meddling in their affairs. We as Lean Hybrid Project Delivery Professionals must be aware that this line exists and exert every effort to not cross that line. We have asked these team members to make commitments to each other not to us, therefore, our purpose in coaching these team members must be clearly understood by both them and us that we are facilitating not micromanaging.

Through Lean Hybrid Project Delivery, we are more nearly able to bring planning into an accurate assessment of the workflow and effort that will be required to accomplish any given goal. This **make ready planning**, is the phase where we begin to learn what each of those individuals actually does in their workspace to make the entire team work.

As this phase is being implemented, you may find it is worthwhile to begin both a constraint log identifying those constraints encountered during this make ready planning and a risk register, identifying risk previously unidentified.

This constraint log is an important learning tool as we plan for the future because we will run into similar constraints in similar work

Lean Hybrid Project Delivery

efforts in future projects. Identifying and logging these constraints will enable us to better pull plan in future projects. Additionally, constraints serve as clues to unseen linkages, or overlooked hand-offs, between the projects that we are working on and other projects that other project managers may also be working on.

Similarly, risks are monitored and managed at the closest level to their probable occurrence, again at the line manger level.

The intention of this system of project management and the fundamentals of the practices involved are to produce predictable uninterrupted workflow. By creating a comprehensible set of commitments that connect the work of the specialists to the promises of the project to the client, this methodology coordinates interactions toward success. (Seed, 2017)

Pro-Tip: Know where you are
Everything associated with Milestone Planning and Pull Planning answers the question – "What SHOULD we do?"

In Make Ready Planning and Local Kanban - we are answering the question – "What CAN we do?"

In the next section, Weekly Work Planning, we are answering the question – "What WILL we do?"

Tool: Weekly Work Planning
Using the same Kanban as the tool described above; weekly work planning is handled at the local level as the team meets weekly at the board to take ownership of the tasks added to the board and move them into the Doing column. This can also be the time to move the done activities. This process informs the line manager who can then cross off the completed activities in the weekly or by-weekly project team meeting. Every weekly meeting at the local work level, at the Gemba, should end with a constant improvement exercise.

Lean Hybrid Project Delivery

Method: Line Managers should collaborate as well
Similar to agile project methodology our adaptation of Lean Hybrid Project Delivery reduced the deliverable milestones to individual work unit deliverables. This enables teams of professionals to work in concert with each other and other workloads to accomplish individual tasks that contribute to the greater project outcome. Until we implemented this methodology, Project Managers were relegated to the wasted efforts of attempting to understand the professional needs of every discipline and specialty that business environment. This waste was compounded by the time and effort consumed in the back-and-forth discovery that the Project Manager was performing to find what work needed to be performed by whom and when. Even as I write the above sentence, I realize how totally unattainable that objective is.

There is no way that any Project Management Professional can assimilate, learn, and understand all the facets of all the professionals that are supporting all the specialties and all the professionals in today's complex business environment. Even the most advanced medical doctors have given up trying to be all things to all people. We as Lean Hybrid Project Delivery Professionals must also realize that we cannot be all things to all people. However, we can coach teams of people who, together can more nearly assimilate the needs of the entire healthcare organization.

Referring back to the Conditions of Satisfaction we will work intentionally with the last planners to help determine what work is needed to accomplish the goal and produce the committed outcome in the duration of time committed to.

Pro-Tip: Include the technical folks
The most difficult challenge that any technical person ever encounters is we tend to "over think" things. **Thinking is what we do for a living**. Much planning is done in the technical person's own

Lean Hybrid Project Delivery

mind. Therefore, this is a challenge to get those technical people to work in a team format and **groupthink**.

Tool: Do the Gemba Walk

The Japanese term for this is Gemba. Gemba, means "the real place" that is generally used to refer to the place **where value is created**. In six Sigma managers are taught to do a "Gemba Walk" which refers to walking about where the work is actually getting done. In Lean processes team areas where technical staff can work together with little or no separation are referred to as a Gemba. In Latin it would be **"locus opus"**.

Lean practitioners through many years have determined that grouping people together by the work they're trying to accomplish, not necessarily by their specialty or discipline, is the most effective way for a team to accomplish an individual task or group of tasks to produce an output product. (Womack J. , 2003)

We as project managers have struggled with this throughout our careers; it will be no different now, but now we have the tools to create the learning experience that will help us overcome this hurdle.

Through the iterations of our conversations in the Lean Hybrid Project Delivery methodology we have experienced improvements in the commitments that are made, timelines that are predicted, cost of producing the outcome, and the quality of the output product. We don't expect our first iterative attempt at this methodology to be perfect, but we do expect to get better with every iteration always moving toward a more perfect outcome.

We are coaching and facilitating to help the DOING be more efficient.

Process: Constant Improvement

People are more important than the processes we use to accomplish the tasks or projects people want to accomplish. Individuals are more

Lean Hybrid Project Delivery

important that the tools we develop. Technology was built for people. People were not invented to make the technology work.

Completed projects are more important than the documentation of the project as it is being done. The **effect** (consequence) of Information Technology is more important than the **affect** (influence) of the project processes of technology as it is being applied to gathering information.

> Practice is everything."
> Periander

The **result** of being able to develop knowledge from the information derived from the data is more important than reporting about how we got there.

Collaboration with the people who work for the business to provide services or products are more important than the management reports about what those people are doing to provide such work product. The knowledge provided by the Information Technology will provide the science to improve work as we know it and benefit all humankind. Reporting is and continues to be an important learning tool, teamwork is the most effective method for improvement.

As opposed to inventing a plan that looks forward, we understand there is a need to be flexible and **embrace change**. We plan more expertly by working retrospectively from the vision's "live perspective" than from the starting point. We plan in more detail, and better detail, as we get closer to the objective. We expect the plans to change, and we expect the path to alter as we move toward a more perfect application of the desired outcome that will better serve the business environment professionals who are in turn building better outcomes for the customer population.

This follows closely the ideology of the Agile Manifesto. (Jeff

Lean Hybrid Project Delivery

Sutherland, 2017)

Knowledge (not Information Technology) is power. The power we seek is the power to improve outcomes for our customers.

We recognize that we will never know less about the processes and steps to accomplish an outcome than at the onset of the project. (Layton, 2012)

Learning as a result of studying the processes and improving on the work we are doing is core to the system of Lean Hybrid Project Delivery. We make every effort to learn from every success and every failure at every step along the way of the project.

Semper Doctrina - Always learning

Learning allows us to take informed actions to **improve** our plans and reduce wasted time and effort in accomplishing our goals. (Deming W. E., 2017)

Our highest priority is to satisfy the Conditions of Satisfaction as early in the process and as efficiently as is possible. We expect changing requirements as we proceed toward the Conditions of Satisfaction because the

> There are four purposes for improvement: Easier, Better, Faster, and Cheaper. These four goals appear in the order of priority."
>
> Shigeo Shingo

Healthcare Industry is always changing and improving. As we learn from and evaluate the work we are doing we are better able to produce elegant outcomes. Elegant outcomes are models of simplicity accented by maximization of the work not done.

Our objective it to eliminate wasted effort and activity in the process of accomplishing our goals. (Agile Alliance, 2017)

Lean Hybrid Project Delivery

We strive to get better and be better at our profession every day, in every iteration of our effort we apply what we have learned and become more efficient and effective.

Questus melius – Getting Better

We are always growing professionally and increasing our competence.

We are always getting better.

Tool: Weekly Check-In Calls:
This meeting is modeled after a Scrum meeting. It is focused, concise, and does not solve problems. Problems are solved outside this meeting, even though this meeting may identify them.
- Is your commitment complete?
- If not when will it be complete?
- Is there a roadblock that is keeping it from being complete?

This meeting is driven from the commitment log.

15 Minute Phone Call

This is not about what you did wrong! It's about how the team can help get it done!

Tool: Bi-Weekly (or Monthly) Big Room Meetings
- Bring the pull-plan back and revisit the activities.
- Mark off the sticky notes that have been completed to date.
- Pull plan new milestones and discoveries.
- Revisit schedules and commitments.
- Apply the Percent Project Complete.
- Perform the plus – delta exercise.

Always. Every step of the way, every day, every week, every meeting, and always. Lean Hybrid Project Delivery affects all aspects of our career.

Lean Hybrid Project Delivery

Tool: Plus / Delta
In the spirit of the Lean Principle of Constant Improvement, never end a meeting without performing a Plus / Delta. This is a brief activity that asks:
- "what did we do well?" and
- "what could we have done better?"

This practice applies to the work completed in the week before as well as to the meeting itself. Write the comments down on a white board and log them.

In the next meeting review. Did we make the improvements we identified las meeting? If not, why? Does it still apply?

Tool: A3 Reporting
We, however, are still obligated to report status to those who are observing our efforts. There is another Toyota Production System model that we'd like to introduce. This is known as the "A3". The A3 is so named because it is on A3 size paper – 11" X 17" (US). John Shook's book "Managing to Learn" is a good read, it breaks it down and illustrates its use. One can also get sufficient information by performing a quick Internet search of "A3 Problem Solving Process".

To summarize the A3 breaks down a report on a single page that is formatted to cover the Deming PDSA Wheel. Reporting is recommended to be visual analytics as opposed to words and numbers. Toyota posts these on the wall or bulletin board. The idea is not that different that the "grease board" of days gone by. Of course, we now use electronic status boards in their place. As a Lean Hybrid Project Delivery Professional, we would like to see these electronic A3s be used as KANBAN boards showing the status updates of the work our teams are performing. (Shook, 2008)

The A3 also captures the budget and cost performance visual

Lean Hybrid Project Delivery

representations. While I have been writing this book the Project Management Institute has published PMBOK Version 7. My copy is on the dining room table, but I did download my free member's copy and reviewed it. If you are in an environment where your work includes monitoring the spend of the project, or if you have a project that is governed by a Purchase Order limitation or other monetary limitation, the PMI has published a number of mathematical formula to assist in calculating the financials. Most Project Managers I have talked to don't use those calculations because the work they do tends to be contract driven. Where this does apply is generally in the areas of construction, infrastructure, and data centers, where I have found a simple burn-down chart makes a good visual representation of the situation. Burn-down charts work well in the A3 methodology of reporting.

Process: Testing and Validation
Testing and validation should be built in at every applicable location in the project. If a milestone has been accomplished, we should be asking is there some part of this product that can be tested and validated to assure the quality expected is achieved. Do not wait until the end to assure quality is as expected. Work collaboratively with the stakeholders and the team to determine if the testing plan is valid and the test results are valid. One should be looking toward the end business environment from the Conditions of Satisfaction to the final delivery of the project.

Process: Mark Tasks Finished
As we discussed earlier, we work hard to chunk down the activities and tasks into work segments that can be completed by and individual or small team in 5 days or less. The reason for this comes as we finish tasks. When you are standing in front of your master Kanban board, if every sticky represents 5 days or less of work it is easy to

Lean Hybrid Project Delivery

calculate (1) how much total work effort there is – how many person hours, (2) as they are completed and marked off – how much work has been done already. Then, by simple division you can accurately assess what percentage complete (by work effort) you have. This number can be translated into velocity as you move through the project.

Pro-Tip: Leverage Vendor Project Managers
In large projects it is often advisable to leverage vendor provided Project Managers.

Tool: Leverage Organizational Change Management Process
To the greatest extent possible, use the established Change Management Process. It will usually include a method for including the appropriate stakeholders. If you cannot do this, assure that you have addressed this in the Conditions of Satisfaction meeting so there will be a known and accepted method for addressing and approving changes that could affect Scope, Schedule, and / or Budget.

Process: Watch Scope Always!
Monitoring and controlling scope is integral to the project. Scope creep kills projects, ruins your reputation, and could cause catastrophic problems. This and risk are always on your mind.

Process: Monitor Risk Always!
Monitoring and controlling risk is integral to the project. Discuss it in every group meeting. Assure you are aware of every potential risk and have a plan for remediating it.

Process: Monitor Transition to Operations
The purpose of the project is to deliver an outcome, a product, or a service. Everything you are doing from day one is moving toward the

Lean Hybrid Project Delivery

successful delivery of that outcome. Be aware of the Outcome Business Environment, revisit it frequently with the key stakeholders and operational staff. Be sure to include the people who will be using it, whether it is a new office building, new hospital, or new software. Include the stakeholders.

Tool: Management by walking about
Different from doing a Gemba walk where you are reviewing the work where it is getting done. This is a process of relationship building, of networking, of strengthening and building trust and friendship. Your being there can often times signal you are a servant leader not a 'task master'.

Process: Lean In
Always be looking for ways to:
- Reduce Defects
- Reduce Rework or over production
- Reduce time spent waiting, this applies particularly to hand-offs.
- Use the talent of the team members at their highest level and develop talents during the project
- Reduce transportation times everywhere possible both on the job site and getting materials into the jobsite
- Reduce stored inventory as much as possible
- Save steps wherever possible – bring the material needed as close as possible to where it is needed
- Eliminate extra processes. In the team meeting be looking for teams or team members who are doing duplicate work.

Focus on workflow and throughput.

Lean Hybrid Project Delivery

9 – Completing the Steps

Completing the Steps

If it is worth doing – it is worth doing excellently.
Each of the processes, methods, and tools outlined in this book has a purpose in successful delivery of your project. It is important therefore at the conclusion of each of these steps that a process be in place that formally closes that step as we move on through the project delivery.

"Good business leaders create a vision, articulate the vision, passionately own the vision, and relentlessly drive it to completion." – Jack Welsh

As with Monitoring and Controlling, in the previous chapter, completing the steps as we pass through this project delivery methodology is an integral part of the process of completing and delivering the product of the project.

Therefore, as we are creating the Conditions of Satisfaction, we are also developing the formal document that is the guiding definition of done. The Conditions of Satisfaction inform the Charette and set the "True North" to guide the entire project.

As we are performing the Charette, we are also further defining the processes of the project by mapping milestones that clearly define our work effort. We are doing this through a process of brainstorming collaboratively with the primary stakeholders so that we are more accurately describing the definition of done. Completing the step of the Charette enables us to better define that part of the project that guides the work effort from week to week as we move forward. Key to this process is those milestones that we

Lean Hybrid Project Delivery

are identifying, the stakeholders then we are identifying, the risks that we are identifying, the level of involvement of the stakeholders that we are identifying, and finally a thorough current state review. At the end of the Charette we have a clear definition at a high level of the desired outcome of the project which was provided in the Conditions of Satisfaction as well as a clear vision of how we will be accomplishing that goal.

Also, beginning in the Charette, we are defining and beginning to understand what the finished business environment will be. From the beginning of the Charette through the end of the project we will constantly be working to transition the project into operations. Therefore, it is important that the users of the final product are included in the Charette and all in the following steps of this project delivery methodology.

Everything in the Charette informs the Flesh-Out. The Charette is revisited as needed and can be performed in mini-Charettes to keep the project workflow efficient.

At the completion of the Charette you should have:

1. Either on a whiteboard or on a 25" by 30" wall sticky - a bullet point list for the Conditions of Satisfaction.

2. Either on a whiteboard or on a 25" by 30" wall sticky – a bullet point list of the key Current State elements.

3. Either on a whiteboard or on a 25" by 30" wall sticky – a bullet point list of known risks.

4. A Kanban board with the Milestones of the project across the top as well as known constraints such as holidays and other

Lean Hybrid Project Delivery

universally constraining dates and elements.

Always perform a plus / delta analysis of the meeting at the end of the meeting.

Completing the Flesh-Out
It is important to understand that the Flesh-Out and the Sequencing can be done in the same set of meetings (if you so desire) and if the project merits that the combination of these two would be more efficient. However, both the Flesh-Out and the Sequencing must be done to be successful. In multiple capital projects, I have determined that it is best not to confine the members during the process of Flesh-Out to having a necessity of sequencing the activities.

The Flesh-Out is a <u>pull planning session</u>. It is highly collaborative and therefore includes many of the line managers as well as their directors. One should include as many of the vendors and contractors that are contributing as well. This is better performed in person as it is high touch - low tech - everybody talking through the process of planning.

At the completion of the Flesh-Out you should have:

1. Either on a whiteboard or on a 25" by 30" wall sticky - a bullet point list for the Test Strategy.

2. Either on a whiteboard or on a 25" by 30" wall sticky – a bullet point list of the Integration Strategy.

3. Either on a whiteboard or on a 25" by 30" wall sticky – a bullet point list of the Future State Review elements.

4. A Kanban board with the Milestones of the project across the

Lean Hybrid Project Delivery

top as well as known constraints such as holidays and other universally constraining dates and elements from the Charette. Updated to include 4" X 4" task level sticky notes for as many of the activities as can be reasonably "pulled" in the planning event. These by nature will be sequenced according to the Milestones but are not yet sequenced by prerequisites.

5. As much as possible each activity should be <u>chunked down</u> to a duration of five day's work effort that can be completed by an individual or small group.

The Flesh-Out can span a meeting or several iterative meetings depending on the availability of the team and the priority of the project.

The Flesh-Out informs the Sequencing meeting. Always perform a plus / delta analysis of the meeting at the end of the meeting.

Completing the Sequencing
It is important to understand that the Flesh-Out and the Sequencing can be done in the same set of meetings (if you so desire) and if the project merits that the combination of these two would be more efficient. However, both the Flesh-Out and the Sequencing must be done to be successful. In multiple capital projects, I have determined that it is best not to confine the members during the process of Flesh-Out to having a necessity of sequencing the activities.

Sequencing is a <u>pull planning session</u>. It is highly collaborative and therefore includes many of the line managers as well as their directors. One should include as many of the vendors and contractors that are contributing as well. This is better performed in person as it is high touch - low tech - everybody talking through the process of planning.

Lean Hybrid Project Delivery

This meeting or set of meetings, depending on the size of the project and the availability of the team members, collaboratively evaluates the activities defined in the Flesh-Out meeting. The view of this pull plan is to associate as accurately as possible the various prerequisites of each of the activities to be completed as well as review and assigning ownership of the activities. The owners should be in participation to the extent possible. This is also where the review of the durations is finalized. Unit testing and integration testing activity stickies should now be included on the Kanban board.

At the completion of the Sequencing you should have:

1. Either on a whiteboard or on a 25" by 30" wall sticky - a bullet point list for the Future State Validation elements.

2. Either on a whiteboard or on a 25" by 30" wall sticky – a bullet point list of the updated risks.

3. Either on a whiteboard or on a 25" by 30" wall sticky – a bullet point list of known unknowns such as expectations of order delays and other supply chain elements.

4. A Kanban board with the Milestones of the project across the top as well as known constraints such as holidays and other universally constraining dates and elements from the Charette. Updated to include 4" X 4" task level sticky notes for as many of the activities as can be reasonably "pulled" in the planning event. These should now be sequenced by prerequisites, have owners assigned, and may be sorted into swim lanes if desired.

5. As much as possible each activity should be <u>chunked down</u> to a duration of five day's work effort that can be completed by an individual or small group.

Lean Hybrid Project Delivery

6. Any activities that are unassigned may be in a "backlog" or "parking lot" panel in the room on the wall to be revisited in a later meeting iteration.

Sequencing can span a meeting or several iterative meetings depending on the availability of the team and both the size of and the priority of the project.

The outcome of Sequencing is:

1. The Risk Register.

2. The initial Commitment Log.

3. A Schedule of Testing Events.

    a. Test Scripts

    b. Workflow diagrams

4. If needed one can now produce an accurate Critical Path.

5. An updated budget.

6. A Communications Plan.

7. Any other documentation required or request by the Project Sponsor or Key Stakeholders.

Sequencing is the end of the formal planning process. Sequencing can be performed as needed to review and adjust the formal completed plan. Substantial changes to the plan should not be allowed after completion of this step, changes should be handled through a formal Change Management process.

Lean Hybrid Project Delivery

The Kanban board should be reviewed weekly to assure timelines and completed activities are identified. The Sequencing informs the Commitment Meetings.

Always perform a plus / delta analysis of the meeting at the end of the meeting.

Completing the Commitment Meetings
Commitment meetings use the commitment log sorted by weeks days within the week and owners of the activities to perform a weekly scrum meeting. The objective of this meeting is clear concise end of short duration. This meeting should not extend longer than 15 to 20 minutes at the maximum. Everyone who has assigned activities for the previous week and for the coming week should be involved in this meeting. In the ideal world everyone would be in one room however in the current state of affairs this can also be handled through virtual meetings such as Zoom.

The facilitator of this meeting is focused on a line-by-line review of the commitment log and produces a update of the commitment log based on whether the project activity is complete, in which case it is marked complete in the commitment log. If the activity is incomplete, then an updated due date is assigned along with a comment as to why the commitment was not met. The objective here is not to assign blame. It is important that the commitment log informs the Kanban board and the remainder of the team effectively allowing the team to adjust to real time delays. The purpose of this meeting is not to fix the problem but to identify it and create a commentary on it. The facilitator may ask if there's anything the team can do to help facilitate the process of completing the activity and if that can be answered clearly concisely and quickly then it will be added to the comment associated to that commitment on the

Lean Hybrid Project Delivery

commitment log. The outcome of an incomplete activity in the commitment meeting may result in a pull planning session that falls back in the sequencing group of meetings.

The other thing that is finished in the commitments meeting is a look ahead for a week at the commitments that are planned for the next coming week to assure that those who are assigned feel confident that they will be able to complete their activities on time. There is no discussion in this meeting of budget. Constraints are only updated in comments to individual commitments.

The facilitator should ask if there are any risk updates. Be aware missed commitments can be risks, caused by risks, or causes for alerting up to the Stakeholders.

Any planning that is determined that needs to be done as a result of the commitment meeting his automatically cycled back to the Sequencing scheduled event.

At the completion of a Commitment meeting you should have:

1. An updated and current commitment log.

2. An updated and current Risk Register.

Like a SCRUM Meeting it is not meant to solve problems, design solutions, or address issues. It is a 'report out' of work completed and confirmation of work planned for the immediate next work period. In this discussion I used a time period of a week but that might also be two weeks if that is appropriate. Plus / Delta analysis in this meeting applies to work done and work to be done in the next week (or associated time period).

Lean Hybrid Project Delivery

This is an iterative meeting whose sole purpose is to facilitate track work actually done and about to be done.

Completing Monitoring and Controlling
All the above are part of the process of Monitoring and Controlling. This method is designed to provide the tools that will enable you as a Lean Hybrid Project Delivery professional to move smoothly through the processes. If you follow these processes, the work should remain in scope, complete as scheduled, and be on budget.

If you haven't already picked up on it the theme that permeates all aspects of this project management methodology his constant improvement. The biggest part of monitoring and controlling in this methodology is the plus delta. Constant improvement as we progress through the project not only helps us speed the project along but also helps us control the quality and improve the quality as we produce the product to deliver the project.

there are a couple of things that I want to recommend before I close this chapter out.

First don't be afraid to use punch lists. I have found punch lists to be very helpful in assuring that everything that is in the design is completed. This is particularly important if you have some level of construction going on when there are multiple trades executing within an area it is important to manage by walking about ... checking that the build is happening as desired. It is much easier to move an electrical box before it is wired and certainly before it is sheet rocked in. However, the concept can also be applied to multiple Information Technology teams where the features are being developed by multiple agile teams and being integrated into the project at a later time. I like punch lists.

Lean Hybrid Project Delivery

The next thing is the testing. Keep focused on testing as early as possible, in every step along the way, frequent iterative testing will often prove to be a most valuable tool in your toolbox.

Remember, as you are moving forward, you're moving toward an operational state. In as much as you are testing, the operational staff that will be running this product (whatever it is) is the validation point. Remember to keep the operational staff in your team so that they can inform you of the reality of the situation in an operational state. This includes telephone operators, Information Technology support staff, facilities engineering, housekeeping, site security, and all other operational departments.

Also, if you are delivering a new facility don't forget all the equipment that is coming in. Where will it be stored, who will install it, and where does all the waste go as you are installing.

Do you need to set up a transition command post? The hand off to operations, even though you have had them along for the ride includes the actual people working in the facility not just the managers you have been working with in the process.

Don't overlook training requirements.

10 - Closing the Project

Now the doors are open the building is in use the operational staff has taken over and you're done right? Well maybe not.

This is a good time to sit down with the leads of the team and review the work in the form of a postmortem. What you're looking to do here is a thorough plus / delta evaluation of the entire project. It's time now to develop a "lessons learned" document.

I find the easiest way to do this is either on a white board or on a series of 25" by 30" wall sticky notes making a bulleted list of what we did well and what we could do better. It might be helpful (depending on the size of the project) to assign or recruit a "village scribe" in this project postmortem. This individual can be a project coordinator or starting project manager. This gives them insight into a project that will help them in their career growth. This individual would take the notes on Microsoft OneNote (or in another format suitable to be published) for the consumption of the executive staff, future project managers and future project teams.

This is also when final tests are performed and validation documents signed off to assure that all the products that were installed or built meet the criteria of the project as defined in the conditions of satisfaction.

In a new facility it's important to remember that everything from paper two pencils to brooms to sweep up dust all have to be brought in it's a good thing to be able to bring that to the attention of the owners to help them quantify what they need to order and where they're going to store it as well as how are they going to disperse it.

Finally, there are all those contracts, procurements, the cleanup of waste products, the disposal of over purchases, and all those products and services that must be scheduled for inventory. Then, when all the

Lean Hybrid Project Delivery

documents are signed, all the procurements are closed, and all the owner validations are complete ... you can finally sit back have a beer and celebrate. The last person to finish the project is the project manager. Congratulations.

11 – Visually Speaking

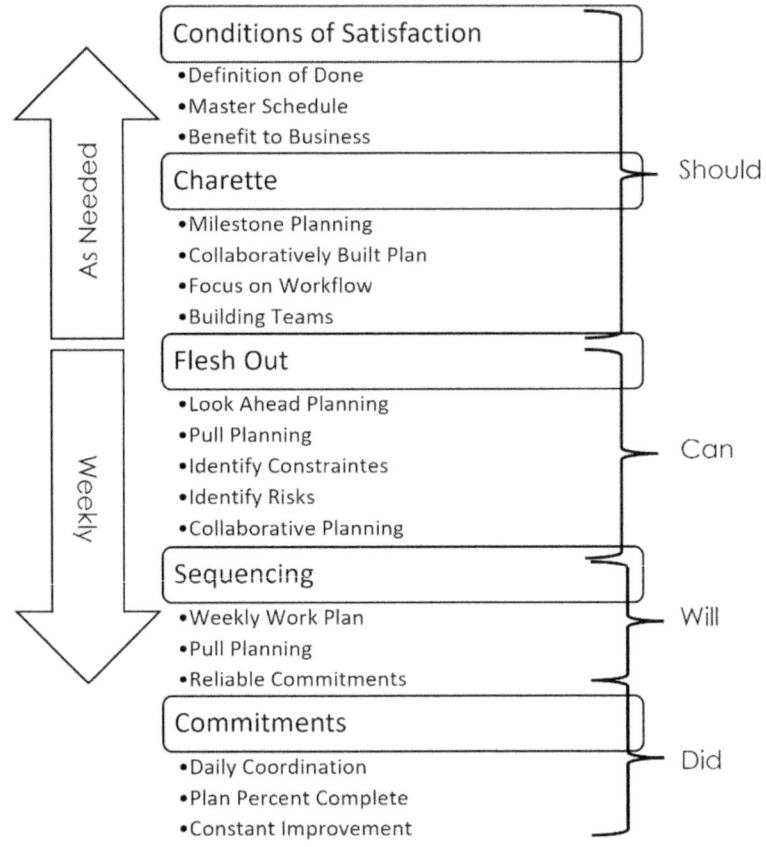

Figure 1 The Lean Hybrid Project Delivery Model

Lean Hybrid Project Delivery

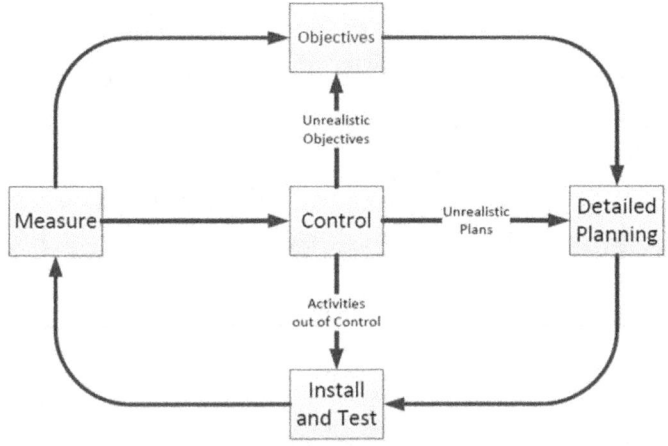

Figure 2 The Monitor and Control Loop

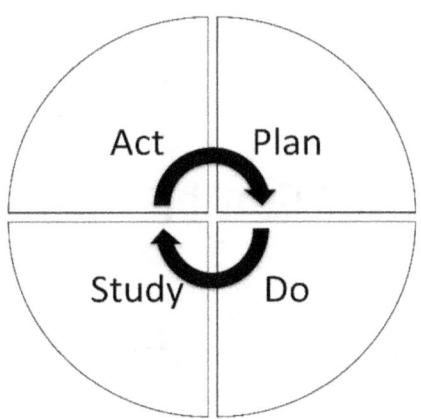

Figure 3 The Deming Model for Constant Improvement

Lean Hybrid Project Delivery

Figure 4 The Plus / Delta Challenge

Figure 5 The Lean Model

Lean Hybrid Project Delivery

Figure 6 The Flores Loop

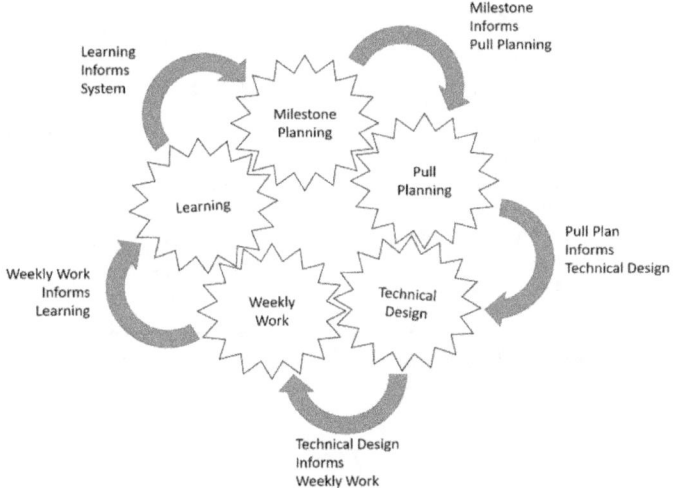

Figure 7 Lean Hybrid Project Weekly Workflow

Lean Hybrid Project Delivery

Figure 8 Pull Planning Objectives

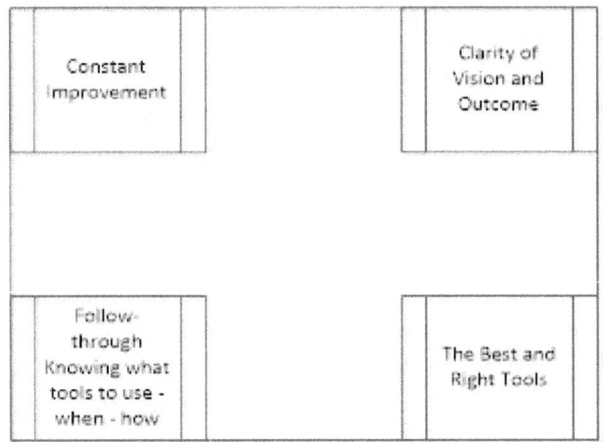

Figure 9 The Cornerstones of Lean Hybrid Project Delivery

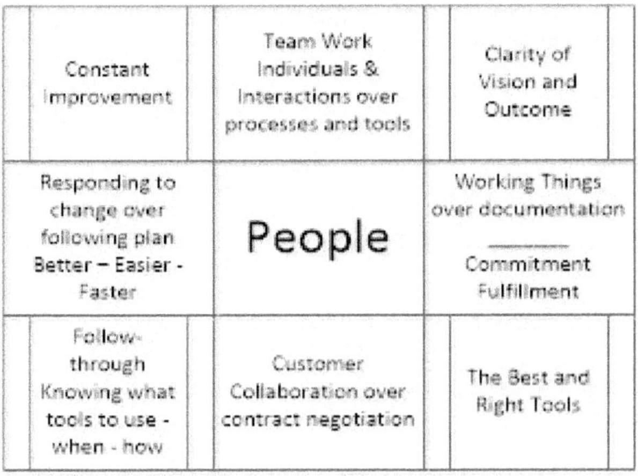

Lean Hybrid Project Delivery

Constant Improvement	Team Work Individuals & Interactions over processes and tools	Clarity of Vision and Outcome
Responding to change over following plan Better – Easier – Faster	People	Working Things over documentation <hr> Commitment Fulfillment
Follow-through Knowing what tools to use - when - how	Customer Collaboration over contract negotiation	The Best and Right Tools

Figure 10 The Agile Foundation of Lean Hybrid Project Delivery

Constant Improvement	Identify Value	
	Pull Planning	Map the value stream
	Create Flow	

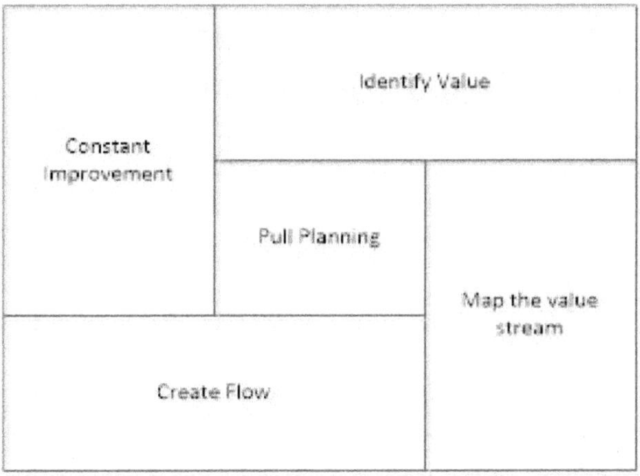

Figure 11 The Lean Layer of Lean Hybrid Project Delivery

Lean Hybrid Project Delivery

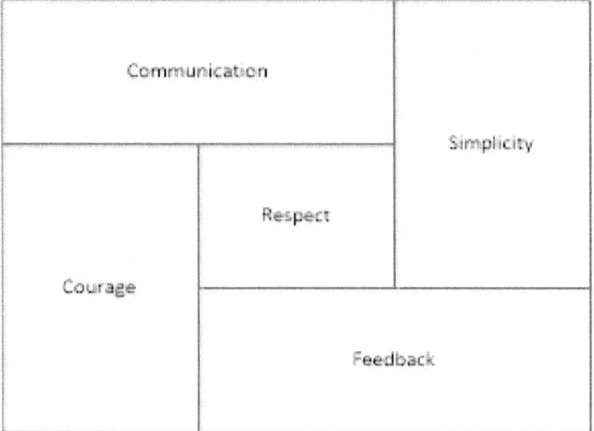

Figure 12 The XP Layer of Lean Hybrid Project Delivery

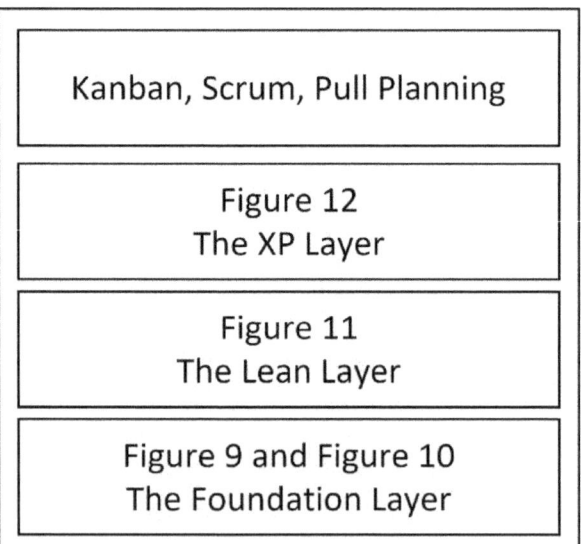

Figure 13 Tying It All Together

Lean Hybrid Project Delivery

Appendices

The Five Lean Principles

1. Specify value from the standpoint of the end customer by product family.
2. Identify all the steps in the value stream for each product family, eliminating whenever possible those steps that do not create value.
3. Make the value-creating steps occur in tight sequence so the product will flow smoothly toward the customer.
4. As flow is introduced, let customers pull value from the next upstream activity.
5. As value is specified, value streams are identified, wasted steps are removed, and flow and pull are introduced, repeat this process again and continue it until a state of perfection is reached in which perfect value is created with no waste.

(Adapted from Womack and Jones 1996, p. 10.)

Lean Hybrid Project Delivery

Manifesto for Agile Software Development

We are uncovering better ways of developing software by doing it and helping others do it.

Through this work we have come to value:

- **Individuals and interactions** over processes and tools
- **Working software** over comprehensive documentation
- **Customer collaboration** over contract negotiation
- **Responding to change** over following a plan

That is, while there is value in the items on the right, we value the items on the left more.

Copyright • Kent Beck • Mike Beedle • Arie van Bennekum • Alistair Cockburn • Ward Cunningham • Martin Fowler • James Grenning • Jim Highsmith • Andrew Hunt • Ron Jeffries • Jon Kern • Brian Marick • Robert C. Martin • Steve Mellor • Ken Schwaber • Jeff Sutherland • Dave Thomas •

Lean Hybrid Project Delivery

The XP Values

Communication
Software development is inherently a team sport that relies on communication to transfer knowledge from one team member to everyone else on the team. XP stresses the importance of the appropriate kind of communication – face to face discussion with the aid of a white board or other drawing mechanism.

Simplicity
Simplicity means "what is the simplest thing that will work?" The purpose of this is to avoid waste and do only absolutely necessary things such as keep the design of the system as simple as possible so that it is easier to maintain, support, and revise. Simplicity also means address only the requirements that you know about; don't try to predict the future.

Feedback
Through constant feedback about their previous efforts, teams can identify areas for improvement and revise their practices. Feedback also supports simple design. Your team builds something, gathers feedback on your design and implementation, and then adjust your product going forward.

Courage
Kent Beck defined courage as "effective action in the face of fear" (Extreme Programming Explained P. 20). This definition shows a preference for action based on other principles so that the results aren't harmful to the team. You need courage to raise organizational issues that reduce your team's effectiveness. You need courage to stop doing something that doesn't work and try something else. You need courage to accept and act on feedback, even when it's difficult to accept.

Respect
The members of your team need to respect each other in order to communicate with each other, provide and accept feedback that honors your relationship, and to work together to identify simple designs and solutions.

Lean Hybrid Project Delivery

Meeting Tips

Weekly (bi-weekly) Meetings:
1. Bring the Pull Plan back
2. Mark off what has been completed to date
3. Review plus delta
    a. What did we do well in our work this past week?
    b. What can we do better in the week to come?
4. Pull Plan new milestones as discoveries (as needed)
5. Calculate planning and effectiveness
    a. Activity stickies marked complete divided by total activities

Sample Meeting Agenda:
1. Ice breaker – 15 Minutes
2. Hot Topics – things happening right now! – 30 Minutes
3. Review Commitment Log – 15 Minutes
4. Coding and customization discussions – 60 to 90 Minutes
5. Six Weeks Look Ahead – Pull Planning – 30 Minutes
6. New Business – 10 Minutes
7. Next Meeting Agenda – 10 Minutes
    a. Pick the next meeting Facilitator
    b. Set location
8. Plus / Delta for the meeting
9. Adjourn

References

Agile Alliance. (2017). *https://www.agilealliance.org/agile101/the-agile-manifesto/*. Retrieved from Agile Alliance: www.agilealliance.org

Alliance, A. (2021). *agilealliance.org/glossary/kanban*. Retrieved from Agile Alliance: https://www.agilealliance.org/glossary/kanban

Alliance, A. (2021, July 27). *Definition of Done*. Retrieved from Agile Alliance: https://www.agilealliance.org/glossary/definition-of-done/

Ammer, C. (2003). *The American Heritage Dictionary of Idioms*. Houghton Mifflin Company.

Andrea Caccamese, D. B. (2012). Beyond the Iron Triangle; year zero. *Paper presented at PMI® Global Congress 2012—EMEA, Marsailles, France*. Newtown Square, PA: Project Management Institute.

Anita Tucker, S. J. (2013). The Effectiveness of Management-By_Walking-Around: A Randomized Field Study. *Harvard Business School*. Retrieved from https://www.mindtools.com/pages/article/newTMM_72.htm

Baki, M. (1998). CPM scheduling and its use in today's construction industry. *Project Management Journal, 29*(1), 7-9.

Ballard, H. G. (2000). *The Last Planner System of Production Contrl*. Birmingham, UK: School of Civil Engineering - Faculty of Engineering - The University of Birmingham.

Ballard, I. D. (2016). *Target Value Design - Introduction, Framework, and Current Benchmark*. Arlington, VA: Lean Construction Institute.

Brechner, E. (2015). *Agile Project Management with Kanban*. Redmond, Washington: Microsoft Press.

Cerner. (2017). *Cerner*. Retrieved from Cerner: www.cerner.com

Davidson, D. (June 22, 2015). *Last Planner System Business Process Standard and Guidelines*. Phoenix, AZ: LCI Israel.

Deming, E. (1994, Nov 19). *The New Economics*. Cambridge, MA: MIT Center For Advanced Educational Services. Retrieved from The Deming Institute: https://deming.org/explore/so-pk

Deming, W. E. (2017, November 18). *PDSA Cycle*. Retrieved from The Deminigs Institute: https://deming.org/explore/p-d-s-a

Dimitris Bertsimas, D. N. (2001). The Distributional Little's Law and Its Applications. *Massachusettes Institute of Technology*, 298-310.

Dodson, C. L. (1865). *Alice's Adventures in Wonderland*. London: Macmillan &

Co.

Dredze, M. (2012). How Social Media Will Change Public Health. *Computer*, 1572-1672.

Flávio Copola Azenha, D. A. (2021). The Role and Characteristics of Hybrid Approaches to Project Management in the Development of Technology-Based Products and Services. *Project Management Journal, 52*, 90-110. doi:https://DOI:10.1177/8756972820956884

Flores, F. (2012). *Conversations For Action and Collected Essays.* North Charleston, SC: Createspace Independent Publishing.

Fowler, S. (2014, Nov 26). *What Maslow's Hierarchy Won't Tell You About Motivation.* Retrieved from Harvard Business Review: https://hbr.org/2014/11/what-maslows-hierarchy-wont-tell-you-about-motivation

Freud, S. (1895). *Studies in Hysteria.*

Gail Lindsey, J. A. (2009). *A Handbook for Planning and Conducting Charettes for High-Performance Projects* (Vols. NREL/BK-710-33425). Washington, D.C.: U.S. Department of Energy Office of Energy Efficiency and Renewable Energy.

Glenn Ballard, I. T. (2016). *Target Value Delivery, Practitioner Guidebook to Implementation.* (K. C. Kristin Hill, Ed.) Arlington, VA: Lean Construction Institute.

Goleman, D. (1995). *Emotional Intelligence.* New York, NY: Bantam Books.

Greaves, M. (Aug 2018). A casual mechanism for childhood acute lymphoblastic leukemia. *Nature Reviews*, 471-484.

Greggory Howell, H. M. (2005). *The Last Planner System: Conversations that Design and Activate The Network of Commitments.* Arlington, VA: Lean Project Consulting.

Harvard Business Review. (2012). *Guide to Project Management.* Boston, MA: Harvard Business Review Press.

Hayes, D. S. (2000). Evaluation and application of a project charter template to improve the project planning process. *Project Management Journal*(31(1)), 14-23.

Heather Craig, B. (2021, Sept 13). *10 Ways To Build Trust in a Relationship.* Retrieved from Positive Phychology: https://positivepsychology.com/build-trust/

Herzberg, F. (1973). *Work and the Nature of Man.* New York: New American Library.

Jeff Sutherland, E. A. (2017). *Manifesto for Agile Software Development.* Retrieved from Agile Manifesto : http://agilemanifesto.org/

Jones, J. W. (2003). *Lean Thinking.* New York: Free Press.

Joomis, D. M. (2007). *Building Teachers: A Constructivist Approach to Introducing Education.* Belmont, CA: Wadsworth.

Kinlen, L. (Dec 1988). Evidence for an infective cause of childhood leukemia. *The Lancet*, 1323-1327.

Krafcik, J. F. (1988). *Triumph of the Lean Production Systems.* Cambridge, MA: MIT Sloan Management Review.

Kristen Hill, C. C. (2016). *Target Value Design - Practitioner Guidebook to Implementation.* Arlington, VA: Lean Construction Institute.

Kyle Kurpinski, T. J. (January 2014). Mastering Translational Medicine: Interdisciplinary Education for a New Generation. *Science*, Vol 6 Issue 218.

Layton, M. C. (2012). *Agile Project Management - for Dummies.* Hoboken, NJ: John Wiley & Sons, Inc.

LCI. (2017, Sept 13). *Glossary.* Retrieved from Lean Construction Institute: https://www.leanconstruction.org/learning/education/glossary/

LCI. (2017, October). *Lean Construction Institute, Lean Articles.* Retrieved from leanconstruction.org: www.leanconstruction.org/learning/lean-articles/

Lean Enterprise Institute. (2017, August). *Lean Lexicon 5th Edition.* Cambridge, MA: Lean Enterprise Institute. Retrieved from Lean Enterprise Institute: https://www.lean.org/search/?sc=true+north

Levy, F. (2019, March). Artificial Intelligence: Implications For Business Strategy - Module 3.

Liker, J. K. (2012). *The Toyota Way to Lean Leadership.* New York: McGraw-Hill.

Lindley, R. (1966, Aug). Recoding as a function of chunking and meaningfulness. *Psychonomic Science, Volume 6*(Issue 8), 393-394. Retrieved from Wikipedia.

Mark Lines, S. A. (2020). *Introduction to Disciplined Agile Delivery - Second Edition.* Newtown Square, PA: Project Management Institute.

Marusak, H. (2019, July 24). Understanding the Psychological Effects of Childhood Cancer. *Scientific American.*

Maslow, A. H. (1943). A Theory of Human Motivation. *Psychological Review, 50(4)*, 370-396. doi:10.1037/h0054346

McChrystal, S. (2015). *Team of Teams*. New York: Penguin.

Merriam-Webster. (2021). *Dictionary*. Retrieved from Marriam-Webster: https://www.merriam-webster.com/dictionary/flesh%20out

Nadkarni, P. M., Ohno-Machado, L., & Chapman, W. W. (2011). Natural language processing: an introduction. *Journal of American Medical Informatics Association, 18*, 544-551. doi:doi:10.1136/amiajnl-2011-000464

NASA. (2017, Sep 13). *The Lean Project Delivery System - An Introduction*. Retrieved from NASA.gov: https://www.nasa.gov/pdf/293166main_56397main_gregory_howell_forum4.pdf

Newport, C. (2016). *Deep Work*. New York, NY: Grand Central Publishing.

Ohno, T. (1988). *Toyota Production System: Beyond Large-Scale Production*. Portland, Oregon: Productivity Press.

Ortiz, V. (2017, Aug 30). *5 Levels of the Last Planner System*. Retrieved from Lean Construction Blog: leanconstructionblog.com/5-Levels-of-the-Last-Planner-System-Should-Can-Will-Did-and-Learn.html

Osborn, A. F. (1953). *Applied Imagination*.

PMI. (2021, Aug 31). *Explore Scope*. Retrieved from PMI.org: https://www.pmi.org/disciplined-agile/inception-goals/explore-initial-scope

Portny, S. E. (2013). *Project Management for Dummies - UK Edition*. Hoboken, NJ: John Wiley & Sons.

Pothitos, A. (2016, October 31). *The History of the Smartphone*. Retrieved from Mobile Industry Review: https://www.mobileindustryreview.com/2016/10/the-history-of-the-smartphone.html

Project Management Institute. (2017). *PMBOK Guide - Sixth Edition*. Newton Square, PA: Project Management Institute.

Richard Toelle, J. W. (1990). From "managing the critical path" to "managing critical activities". *Project Management Journal, 21*(4), 33-37.

Robert Conti, J. J. (2006). The Effects of Lean Production on Worker Stress. *International Journal of Operations & Production Management ·, 26(9)*, 1013-1038. doi: 10.1108/01443570610682616

Russell, S. J. (2003). *Artificial Intelligence: A Modern Approach*. Upper Saddle River, NJ: Prentice-Hall.

Schwalbe, K. (2011). *Information Technology Project Management - Revised 6e*.

Boston, MA: Course Technology.

Seed, E. A. (2017). *Transforming Design and Construction: A Framework For Change.* (W. Seed, Ed.) Arlington, VA: Signature Book Printing (LCI).

Sergey Asvadurov, T. B. (2017, September). *The art of project leadership: Delivering the world's largest projects.* McKinsey and Company, Megaproject Organization, Design and Delivery. London, UK: McKinsey and Company. Retrieved from https://www.mckinsey.com/business-functions/operations/our-insights/the-art-of-project-leadership-delivering-the-worlds-largest-projects

Shook, J. (2008). *Managing to Learn.* Cambridge, MA: Lean Enterprise Institute.

Sloan, M. (2021, September 2). *System Dynamics and Systems Thinking Resource Guide.* Retrieved from MIT Systems Dynamics: http://web.mit.edu/sysdyn/sd-intro/D-4321-6.pdf

Sten Lindahl, F. M. (2018). *Translational medicine.* Encyclopædia Britannica, inc.

Sweis, B. M. (2018). Sensitivity to "sunk costs" in mice, rats, and humans. *Science, 361*(6398), 178-181. doi:10.1126/science.aar8644

Tan, P. (2020, October 14). *Session 6: Agile Project Management.* Retrieved from MIT Open Courseware: https://ocw.mit.edu/courses/comparative-media-studies-writing/cms-611j-creating-video-games-fall-2014/lecture-videos/lecture-6-agile-project-management/

Teresa M. Amabile, S. J. (2011, May). The Power of Small Wins. *HBR Magazine*, p. Reprint R1105C.

Thorsten Litfin, J. M. (2011). How to Introduce Lean Management to Small and Medium-Sized Enterprises. *NTNU Engineering Series No.1: Proceedings of MITIP 2011 - At: Trondheim Volume: 1.* Trondheim, Norway: MITIP 2011. Retrieved from https://www.researchgate.net/publication/271365422_How_to_introduce_lean_management_to_small_and_medium-sized_enterprises_title

Trotter, F. (2013). *Hacking Healthcare.* Sebastopol, CA: O'Reilly Media.

Uhlman, F. T. (2012). *Hacking Healthcare.* Sebastopol, CA: O'Rielly Media.

Walton, M. (1986). *The Deming Management Method.* New York: Perigee.

Lean Hybrid Project Delivery

Walton, M. (1986). *The Deming Management Method*. Penguin Group.

Webster Dictionary. (2017). *Word of the Day*. Retrieved from Merriam-Webster: www.merriam-webster.com/dictionary/vade%20mecum

Winston, P. H. (2018, January). *RES.TLL-005 How to Speak*. Retrieved from Massachusetts Institute of Technology: MIT OpenCourseWare: https://ocw.mit.edu.

Womack, J. P. (2007). *The Machine That Changed The World*. New York, NY: Free Press - Simon and Schuster.

END

www.ingramcontent.com/pod-product-compliance
Lightning Source LLC
Chambersburg PA
CBHW060852170526
45158CB00001B/326